Hypnofacts 5

Trevor Eddolls

This book is dedicated to

Jill, Katy, Harry, Freddy, Jennifer, Andy, Jake, and Rory

First published in 2017

By iTech-Ed Hypnotherapy

16 Brinkworth Close

Chippenham

Wilts SN14 0TL

Typeset by iTech-Ed Ltd

The right of Trevor Eddolls to be identified as the author of this work

has been asserted in accordance with Section 77 of

The Copyright, Designs, and Patents Act 1988

978-0-244-34092-6

Contents

Introduction

Like its predecessors, this book also contains various articles for hypnotherapists covering practical issues such as working with clients with anxiety, working with children, and helping people deal with issues around social media, checking e-mails, and nomophobia. There are some example word patterns and ideas for using Appreciative Inquiry in the talking part of the session. And there are more theoretical issues such as Socratic questioning, clinic culture, and understanding how GDPR affects practitioners.

Again, the articles assume a model of the brain in which core activities (such as telling the heart to beat) are handled by the brain stem, more protective functions (such as fighting, fleeing, feeding, and reproductive behaviour) are handled by the primitive emotional brain, and higher functions (such as problem solving, maintaining attention, and controlling emotional impulses from the primitive brain) are handled by the intellectual brain. In terms of physical parts of the brain, these three areas more-or-less match up to the brain stem and cerebellum, the limbic system, and the cerebral cortex. It also assumes that the primitive emotional brain is very fast and the intellectual brain is much slower and tends to be used less.

In addition, the book assumes that the mind and body make up a single functioning system that is affected by each other and the environment they are in.

And it assumes a solution-focused model for hypnotherapy – moving clients towards their desired goals rather than worrying about the problem itself and its origin.

Accepting others

A look at strategies to help a client accept other people's behaviour.

My client, let's call her Annabel, arrived feeling harassed and exasperated. A few deep breaths later, she was able to explain that she had recently moved in with a new partner. The household included her two children (who she described as fairly normally horrible) and her new partner's two children (who she described as rude and selfish). I inferred that she wasn't the most comfortable person around children – and she later confirmed this more than once.

What she wanted was help being able to accepts her partner's children's rudeness, and to not get het up (her words) about it every time. In fact, she was now getting 'het up' at the expectation of having to interact with them and was finding going home to be difficult.

I described the brain to her and she told me that she wasn't much of a brooder, so she didn't replay the events in her head over and over. It's just when the events happened, she hoped her partner would intervene and say something appropriate, but he never did. And she didn't feel it was appropriate for her to explain to his children what she considered was an appropriate way for them to act. So, she concluded that she needed help to be more accepting.

I asked her a little bit about her lifestyle. She explained that she was an executive at a medium-sized company and was very busy. She would often take work to bed and sleep once it was finished. There wasn't a lot of time for relaxation in her life.

In the first session, I recapped on how relaxation helped us to get back into using our intellectual brain and make logical decisions about how we wanted to respond to events. I reminded her what Charles R Swindoll said that life is 10% what happens to me and 90% of how I react to it. Or Epictetus, the Stoic philosopher who was a Greek-born slave in Rome in the first century, who said: "It's not what happens to you, but how you react to it that matters".

I also reminded her to give herself permission to relax and spent the time in trance being in a relaxed state (Rag doll as a induction/deepener, and Float away stress as the main script with a little bit of confidence).

In session two, we again talked about relaxing and getting into the intellectual brain for that logical control. And we talked about bucket emptying and being less prone to overreact to events – because the fast pathways are in the primitive brain, with the intellectual brain working much slower.

I also spent time giving her some techniques to help her relax when facing 'the enemy' (again, her words). We tried 7-11 breathing. Then I showed her the peripheral vision technique for relaxation. And, lastly, I anchored the feeling of being calm and in control. These were techniques she could take away and use immediately.

In the next session, we looked at all the positive qualities of Annabel's partner's children. A sort of Appreciative Inquiry technique and one that I knew she was familiar with from her job. And in the session after that, we looked at some of the CBT cognitive distortions, such as 'all or nothing' thinking, catastrophizing, emotional reasoning, and 'shoulds' and musts'. We also looked at other (positive) ways the children's behaviour could be interpreted. And we looked at self-fulfilling prophecies, where if you expect something to turn out badly, it does!

> Sociologist Robert K Merton coined the expression "self-fulfilling prophecy". A self-fulfilling prophecy is a prediction that directly or indirectly causes itself to become true, by the very terms of the prophecy itself, due to positive feedback between belief and behaviour.

In the following session, after recapping etc, we looked at ways that she might be able to turn off her anger when she could feel it beginning. These techniques included:

- Breathing
- Laughing
- Visualizing a relaxing time
- Using a calming phrase (eg saying 'relax')
- Exercising before meeting them
- Making a physical change such as walking to a sunny window
- Counting to 10 before responding.

All the time I was helping her to empty her bucket and stay in her intellectual brain. In the next session, we did a short Mindfulness relaxation exercise, where all she did was focus on her breathing for three or four minutes. A technique that she could take away with her and use at any time. We also looked at the Mindfulness technique of urge surfing and how she could surf the urge (rather than repress the urge) to respond to the children's behaviour.

In the following session, we talked about managing her expectations – so she didn't expect the children to behave 'normally' (her definition of normal, anyway), and accepted their behaviour as the way they behaved now. This separated their behaviour from the way they might behave in the future. And I convinced her to replace sentences with 'shoulds' and 'musts' in them with the phrase, "I would prefer it if…".

At a later session, we did discuss how she might try to change the children's behaviour. We talked about separating the person from the behaviour and using the format, "when you (do whatever), I feel (whatever)". We talked about the 'U-shaped' conversation where you start with a compliment, then move on the criticism, and end with a positive statement. Viktor Frankl said: "Between what happens to us, the stimulus, and how we respond to that stimulus, there is a space, and in that space is our power to choose our response. In our response lies our growth and our freedom".

In another session, we looked at some things to think about so that Annabel would be more relaxed with her partner's children and stop trying to change their behaviour. We considered:

- Why her view of how children should behave was right and how she could accept her partner's children as they are.

- How to respect their individuality, they're complex mixtures of opinions, emotions, values, knowledge, experience, dreams, interests, and goals.
- Their good points
- What it would be like for Annabel to be one of them.
- Being compassionate.
- Being grateful for having them in her life and the things they brought to her life.
- Accepting them and everything they do.
- Stopping comparing them to some ideal.

The sessions were making big difference to how Annabel was feeling and acting between sessions. And she did have a conversation with her partner about how his children behaved. After that, she concluded that the best way forward was to accept their behaviour as part of the give-and-take of living with someone. Being a very determined person, she was able to turn down the dial in her brain so that what would have upset her previously would now be noticed and quietly ignored. A positive result for everyone in the household.

References:

http://www.positivelypresent.com/2011/11/what-if-you-accepted-people-just-as-they-are.html

https://www.howtoforgivepeople.com/how-to-accept-things

http://www.huffingtonpost.com/daylle-deanna-schwartz/accepting-the-reality-of-_b_5052371.html

http://www.purposefairy.com/710/challenge-the-behaviour-not-the-person/

https://www.pickthebrain.com/blog/9-ways-stop-trying-change-people/

Anxiety

A look at the different types and causes of anxiety.

Anxiety can be defined as feelings of unease, worry, and fear that can be mild or severe. Anxiety refers to both the emotions and the physical sensations a person might experience when they are worried or nervous about something.

The first thing to understand is that anxiety is created in two different parts of the brain. There's the amygdala type of anxiety, where you're staying alert and jumping at any dark shadow. And there's the prefrontal cortex type of anxiety, where you start to worry about whether you left the hob on, or you left the cat in/out, or whether you locked the back door before you left by the front door.

Distressing thoughts are more likely to come from the left hemisphere because logical reasoning occurs in the left hemisphere. Rumination or brooding is where a person repetitively mulls over an idea. Rather than coming up with an answer to the problem, this continued dwelling on a problem strengthens the circuitry leading to anxiety (neurons that fire together, wire together – Donald Hebb.) The right hemisphere creates anxiety based on imagination and visualization. The amygdala can become highly activated when the right hemisphere creates frightening images. Vigilance – a general state of alertness – is also based in the right hemisphere. Your cortex may see something – an event – like ambulances dashing down the road. It will then interpret that event – someone at my house has been taken ill – and this will lead to the emotion of anxiety in the amygdala. It's the cortex's ability to predict future events that gives us this ability to feel anxious.

The left prefrontal cortex is where a person plans and executes actions. We can anticipate events positively or negatively. The left prefrontal cortex is where anxious apprehension comes from. If a person finds themselves imagining frightening scenarios, those scenarios are being produced by the right prefrontal cortex.

Your amygdala is stimulated to be anxious when messages arrive from the thalamus indicating that there may be some danger. It will then kick off the HPA (Hypothalamus, Pituitary, Adrenal) axis – taking the body into fight or flight mode. And if it can't do that, it may use the vagus nerve to produce the more primitive freeze response. Usually, the same messages reach the cortex, which then decides that the threat is no more than a plastic bag (or some other non-threatening item). It then tells the amygdala to stand down.

The amygdala is also stimulated by the anxiety messages coming from the cortex. These anxiety messages have been created by the cortex. The amygdala will then also initiate the HPA axis and a person will find themselves feeling very anxious without there being a definite cause, and nothing to fight against or runaway from.

A stress response looks like:

- Pounding heart
- Rapid breathing/hyperventilation
- Stomach distress/nausea

- Diarrhoea
- Muscle tension
- Wanting to run away
- Perspiration/sweating
- Difficulty focusing
- Immobilization
- Trembling/shaking
- Chills or hot flushes.

> When anxious people keep breathing through an open mouth they may hyperventilate, which can cause panicky feelings. And that can make them try to breathe in even more. The secret is to breathe out and then breathe in through the nose and slow down their respiration rate.

There are ways to overcome feelings of panic. If it's coming from the amygdala: try deep breathing, muscle relaxation, and exercise. If it's coming from the cortex: remember it's only a feeling, don't focus on the panic attack, try to distract yourself, and don't worry what other people think.

Being relaxed can reduce feelings of anxiety. Good ways to relax include:

- Slow deep breathing – inhale slowly and deeply, and exhale fully.
- Diaphragmatic breathing (abdominal breathing) – this is thought to massage the internal organs. Place one hand on your chest and the other on your stomach. Take a deep breath. Your stomach should expand.
- Progressive muscle relaxation – this involves tensing and then relaxing one muscle group after another. Start with your hands then up your arms. Next tense and relax your feet, and work your way up your legs. Finally, start at the top of your head and work down your face, into your neck, your shoulders, and stomach.
- Visualizations – we probably know quite a few of these!
- Meditation – you simply concentrate on your breathing. Every time your thoughts move away from your breathing, you bring your focus back to your breath.
- Exercise – this affects the noradrenalin and serotonin levels of the amygdala, making the receptors less active. It also stimulates the left pre-frontal cortex more than the right. This has been associated with a more positive mood (and that helps to reduce anxiety).
- Sleep – getting the right amount of sleep helps people to concentrate and helps them remember things. It also makes the amygdala less reactive.
- Avoid catastrophizing (thinking that everything is currently awful or everything is going to be awful).
- Cognitive defusion – this technique originated with Steven Hayes. Basically a person acknowledges a thought exists without accepting it, eg saying: "ah, once again I'm having a thought about failing my driving test". It's a dissociative technique and allows a person to observe their cortex working.
- Cognitive restructuring – this technique gives you a way to change your cortex. The key is to be sceptical about anxious thoughts and challenge them with evidence,

ignore them, or replace them with new coping thoughts.

- Plan rather than worry – if you're anticipating something bad happening, don't worry about the event happening, plan some solutions. So, if the event does occur, you can execute your plan.

- Engage the left hemisphere – events such as watching comedy programmes, reading articles, playing games, or exercise reduce the dominance of the right hemisphere.

- Engage the right hemisphere – listening to music or singing engages the right hemisphere, so it can't be negative.

Jonathan Haidt in *The Happiness Hypothesis* gave us the elephant and the rider metaphor to represent the intellectual and the primitive brain. Perched on top the elephant, the rider holds the reins and seems to be in charge. He can get the elephant to do lots of heavy work such as lifting logs etc. But if the elephant becomes scared, it will run away with the rider and there's nothing the rider can do about it.

So, just using the intellectual brain to try not to be so anxious won't work, because the primitive brain will run away. The rider and the elephant must be working towards a common goal to make positive changes.

- Mindfulness – this results in the cortex responding to anxiety in a different way. It activates the ventral medial pre-frontal cortex and the anterior cingulate cortex, which are the parts that have a direct connection to the amygdala.

- Cupped hands breathing – putting your hands over your nose and mouth makes you breathe in the air that you've just breathed out. This gets the CO_2 level back to normal, which makes people feel calmer.

- Count backwards – start at 200 and notice which number you get to when your heart rate starts to slow down.

- Use coping statements – say things like, "I know I can stay calm, cool, and in control", or, "I know I feel scared at the moment, but this feeling will pass", or, "I don't like this feeling, but I know that I can cope with it".

Other useful information about anxiety includes:

- A trigger is any stimulus (sensation, object, or event) that becomes associated with an emotional memory of a negative event. Whenever something triggers a response, the amygdala produces a fear reaction and a learned behaviour is initiated. This gets stronger the more times the trigger occurs. The lateral amygdala doesn't look for cause and effect, only association between two events.

- Cognitive fusion is where we assume that what we think is real is actually real. A common example is believing that a situation is dangerous because of a feeling that it's dangerous rather than there being any evidence that there's a threat.

- New learning in the amygdala occurs in the lateral nucleus. This is where you can train your amygdala to respond differently.

- Optimism is more associated with left pre-frontal cortex activation and pessimism is associated with right pre-frontal cortex activation.

- The right hemisphere has a tendency to focus on negative visual or auditory information.

- The more activity there is in the nucleus accumbens (near the limbic system) – an area associated with hope, optimism, and anticipation of rewards and where dopamine is released into – the more dopamine that gets released and the more optimistic a person is. Optimistic people are less anxious.

- Worry arises in the orbitofrontal cortex. This is an area of the brain that allows us to make plans and exhibit self-control. If we focus mainly on negative outcomes, this becomes worry.

- The anterior cingulate cortex can get stuck on certain ideas or images, and this contributes to worry.

In *How Full Is Your Bucket? Positive Strategies for Work and Life* by Tom Rath and Donald O Clifton, Don Clifton came up with the idea of the metaphorical mental bucket. The difference with his idea is that a full bucket is good and an empty bucket is bad; and bad events/thoughts/etc dip into your bucket and empty it.

Reference:

Catherine M Pittman and Elizabeth M Karle Rewire your Anxious Brain: How to Use the Neuroscience of Fear to End Anxiety, Panic and Worry. New Harbinger, ISBN-13: 978-1626251137

Lynda Hudson. Scripts and Strategies in Hypnotherapy with Children (2009). ISBN-13: 978-1845901394

https://en.wikipedia.org/wiki/Nucleus_accumbens

Tech-OCD

A look at how enthusiasm for modern tech can affect some people's mental health.

Human bodies evolved to live successfully on the plains of Africa. Our guts symbiotically worked with bacteria to make the most of the food that was available; our bodies functioned perfectly with frequent movement; and our brains adapted to keep track of around 150 people who were members of our family group/clan/tribe.

But now, with social media, we can spend long periods of the day keeping track of more than 150 people, as we check Facebook to see what our 300 or so friends are doing; look through our Twitter feed at the 200 people we follow on that; and browse LinkedIn at the updates from another group of around, say, 500 people. It's no wonder that we need to keep checking, our brains can't cope with the extra information (and those funny cat videos – called lolcats (look it up!)).

> A client told me that he was addicted to Twitter.
>
> I said that I didn't follow him!!

So, how can you tell whether a client is just browsing through their Facebook feed or whether they're obsessive about it? There are some tell-tale signs to look out for. These include:

- Over-sharing – that's not just sharing some information about their lives, but sharing intimate information. And this may be because they need the gratification of being acknowledged or receiving peer approval. But, it may well be that they are unable to judge what's appropriate to share, with the need to be heard overriding any privacy concerns.

- Reporting on Facebook – ie using Facebook as a log of their every activity, no matter how small or inconsequential. This could be a sign of obsession, as if they need to post something, no matter how ordinary or unimaginative, in order to relieve their anxiety of not doing so.

- Checking Facebook all the time – it may be that they're waiting for a relative to post news of the birth of a new niece or nephew and they're keen to get the news as soon as possible. This might be quite normal if they're a continent away or other valid reasons. But what if they're checking it every few moments at work, or, worse, every few moments when out with friends. Not only are people checking other people's posts but they are also looking at responses to their own posts. Continually checking other people's posts is called FOMO (Fear Of Missing Out).

- Excessive time browsing Facebook every day – while spending some time each day checking a person's newsfeed etc is fairly normal, spending a lot more time each day doing it could mean there's a problem. And if your client is losing sleep time to spend on Facebook, this is going to impact on their whole life.

- Overly concerned with their Facebook image – this is where they ponder for ages before posting an update and then eagerly anticipate others' responses. Consequently, what people think about them on Facebook becomes an all-consuming activity.

- Adding more-and-more friends – this can be an indicator of a Facebook addiction, especially when a person feels that they are in competition with their friends to gain the most friends on Facebook. Research from Edinburgh Napier University found that Facebook users with more friends tended to be more stressed when using Facebook.
- Compromising offline social life – people can feel more comfortable socializing online than offline – in the real world.

Some people even create accounts for their pets, and post updates about what they are doing! But why? What makes people obsessional about social media? One answer seems to be that social media addiction activates the same areas of the brain as drugs such as cocaine. Researchers found the Facebook triggers activated the amygdala, which helps establish the significance of events and emotions, and the striatum, which is involved in the processing and anticipation of rewards. The good news is that the researchers speculated that the addictive behaviour with social media stems from low motivation to control the behaviour, which is due partly to the relatively benign societal and personal consequences of technology overuse, compared to, say, substance abuse.

Another big problem with Facebook is envy. All of a person's friends seem to dress smarter, go to nicer places (on holiday and just for a night out), get to meet 'important' people, and eat fantastic meals compared to them.

So, what do they do? They see a post from their friends that looks fantastic, and they compare their friend's life (or this tiny snapshot of their life) with their own life. And then they feel inadequate. This is followed by feelings of sadness and depression. And, worse, this may be followed by a decision to emulate their friend's life and book a holiday they can ill afford or buy similar clothes or just stop being themselves and become a copy of their friend's imagined life. None of these reactions is likely to end well.

The obvious solution to these issues of envy, anxiety, and depression is to stop using social media – to simply disconnect a person from Facebook, Twitter, Instagram, LinkedIn, Snapchat, WhatsApp, and the rest for a while. But feelings of being disconnected can lead to, what's being called, nomophobia. Nomophobia is the name being given to the fear of having no mobile phone – and this could be from loss, forgetfulness, the battery running out, etc – and, of course, having no mobile phone means a person can't access these social media apps. Long gone are the days when mobile phones were used to just make phone calls!

You won't find nomophobia in ICD-10 (The International Statistical Classification of Diseases and Related Health Problems), which is a medical classification list by the World Health Organization (WHO) containing codes for diseases, signs and symptoms, abnormal findings, complaints, social circumstances, and external causes of injury or diseases. Nor does it occur in the DSM-5 (The Diagnostic and Statistical Manual of Mental Disorders), which acts as a universal authority for psychiatric

Cyberchondria is a health anxiety that can lead to unnecessary medical appointments and tests, and is fuelled by looking up symptoms on the Internet.

diagnoses in the USA. However, it is very likely that it will be included in future editions of both.

People not only use their phones to easily upload photos and comment on social media, they also use them to browse Web sites, listen to streamed music and radio, listen to podcasts, and check their e-mail. Once upon a time Blackberry owners would respond to a ping from their phones every time an e-mail arrived, but now, with so many e-mails arriving, the device would be pinging all the time! All these e-mails means that people get in the habit of checking their e-mail regularly.

It may be that your client is expecting an important message, but if they wake up in the night and check social media and their e-mail, the likelihood is that they have FOMO – the Fear Of Missing Out (mentioned earlier). Another aspect of FOMO can be a reluctance to delete e-mails, just in case they contain some gem of information that a person might need later. It can also lead to having multiple e-mail addresses and checking each of them. This adds social anxiety to our list of problems with using technology.

You may also see clients who spend much of their day huddled over their tablet looking at the statistics being bluetoothed from their Fitbit or other wearable device that counts their steps (pedometer), monitors their pulse rate, and tells them how well they are sleeping. Different devices can measure other aspects of a person's physiology. Wearable tech is becoming commonplace.

There is a suggestion that some of the targets built into these devices may be doing us harm. Many devices recommend 10,000 steps a day, but there's not a lot of evidence that this is right for everyone – and for some people may be harmful. Some devices can calculate every calorie a person takes in and every calorie of energy they use, and this can lead some people to unusual eating habits and even, possibly, to anorexia. Plus there can be added stress if a person isn't getting the required amount of exercise in a day or the correct amount of sleep at night – and that last thing people need is more stress in their life.

Or, you may well see clients who play video games excessively – these are people who sacrifice sleep time and socializing time in order to play a game. It now seems that playing video games can change how a person's brain performs as well as its structure. Video game players display improvements in sustained attention and selective attention. And the regions of the brain that play a role in attention are more efficient in gamers compared with non-gamers. Playing video games apparently increases the size of the right hippocampus, making the player better visuospatially.

On the down side, Internet gaming disorder can be given as a diagnosis for people who are gaming addicts. These people have functional and structural alterations in their neural reward system, which makes them want to continue playing the game (rather than sleep or interact with people in the real world).

You may also see people who aren't addicts and aren't continuously on Facebook, but are struggling with sleep because of their use of modern technology. LEDs (Light-Emitting Diodes) are found in TVs, phones, tablets, kindles, and other popular devices. Our bodies' circadian rhythms control the timing of physiological processes such as sleeping, feeding, hormone production, and cell regeneration. The hypothalamus sets its sleep patterns to match daylight. When it starts to get dark outside, the hypothalamus tells to the body to start making sleep hormones, like melatonin, and to drop the body's temperature ready for sleep. In the morning, when it starts getting light, the body warms up and produces hormones, like cortisol, to wake up. When people are working on their tablets until late into the evening, the body doesn't receive any signs that it's getting dark outside and so its response is much reduced.

The other problem with LED devices is that they produce blue light, which boosts attention, reaction times, and mood – which is not what anyone wants just before bedtime. Blue light also reduces melatonin production more than ordinary light. Blue light also suppresses delta brainwaves (the ones that induce sleep), and boosts alpha waves (creating alertness). All of which make going to sleep more difficult.

What can you recommend to your clients? Firstly, recommend they avoid blue light for at least half an hour (an hour might be better) before they go to sleep. In Windows 10, in Settings, users can turn on the option to 'Lower blue light automatically'

How else can we help clients with technologically-linked issues such as nomophobia, depression, anxiety, addiction, and insomnia? The answer is straightforward – in the usual way. As solution-focused hypnotherapists, we work with clients to attain their goals. And we do that, mainly by emptying their metaphorical stress buckets, helping them to relax, and making positive changes to their behaviour (often this involves helping them to create new habits). It's a case of finding out when the bad stuff doesn't happen and getting them to do more of whatever they do in those circumstances; utilizing their strengths; and celebrating their successes with them. And then they'll be able to reduce the amount of time they spend on social media or online gaming and more time sleeping and interacting with real people.

References:

Social Network Size Linked to Brain Size: https://www.scientificamerican.com/article/social-network-size-linked-brain-size/

7 Telltale Signs of Facebook Addiction: http://www.hongkiat.com/blog/facebook-addiction-signs/

7 Signs of Facebook Addiction: https://www.lifewire.com/signs-of-facebook-addiction-2654371

Facebook addiction 'activates same part of the brain as cocaine': http://www.telegraph.co.uk/news/12161461/Facebook-addiction-activates-same-part-of-the-brain-as-cocaine.html

Facebook Addiction Disorder — The 6 Symptoms of F.A.D.: http://www.adweek.com/digital/facebook-addiction-disorder-the-6-symptoms-of-f-a-d/

6 Stages of Facebook Envy: https://www.daveramsey.com/blog/6-stages-of-facebook-envy

Nomophobia: A Rising Trend in Students: https://www.psychologytoday.com/blog/artificial-maturity/201409/nomophobia-rising-trend-in-students

How to Overcome Email FOMO (Fear of Missing Out): http://www.asianefficiency.com/email-management/overcoming-email-fomo/

Health apps could be doing more harm than good, warn scientists: https://www.theguardian.com/science/2017/feb/21/health-apps-could-be-doing-more-harm-than-good-warn-scientists

How video games affect the brain: http://www.medicalnewstoday.com/articles/318345.php

How Blue LEDs Affect Sleep: https://www.livescience.com/53874-blue-light-sleep.html

Hypnotherapy for weight issues

Here are some suggested ways that hypnotherapy can help clients struggling with weight issues.

Probably, the majority of clients who come through the door with weight issues want to lose weight. Some of them want a simple magic formula that will allow them to continue eating and drinking the same amounts, while, at the same time, losing weight, looking slimmer and younger, and feeling generally fitter. All of them can benefit from solution-focused hypnotherapy.

It's estimated that around 1.6 million people in the UK are affected by an eating disorder, with 14-25 year olds most at risk of developing this type of illness. Women (and girls) are ten times more likely than men (and boys) to suffer from anorexia or bulimia. In addition, there are a number of illnesses associated with eating disorders, such as: type 2 diabetes, high blood pressure, high cholesterol leading to stroke and heart attack, osteoporosis, blindness, and limited life expectancy.

The first thing to do is to establish their goal. How much weight do they want to lose and by when. It can be useful at this stage to produce a roadmap for them showing how much they'd have to lose each week to achieve their goal. But always, with people who are overweight, remember that there may be underlying reasons why they are overweight – and these reasons may well be revealed through the miracle question and can be dealt with in later sessions. And that's why it's important to acknowledge the client's presenting problem, but use the standard techniques to help them overcome other personal issues at the same time.

Having goals and targets can be a great way of motivating clients with high self-esteem. However, clients with low self-esteem are often frightened by goals and fear that they will fail to achieve them – and then make sure that they do fail! So, for those people, there's no point setting weekly targets, although longer-term goals are important.

I always get clients to start a food diary – so they really know how much they are eating and what they are eating. I also talk about exercise and how calories in should be lower than calories out if they want to lose weight. And I also point out that people who eat the majority of their calories later in the day tend to eat more calories, weigh more, and have more body fat. So they may want to eat more of their daily calorie intake for breakfast and lunch rather than dinner or supper. I also recommend that they cut down on take-aways and processed foods and increase their intake of fruit and vegetables.

Typically, between 60% to 85% of the calories you use in a day are for things like breathing, digestion, and circulation. This is your Resting Metabolic Rate (RMR). It's worth noting that digesting the food you've eaten burns calories.

You might think that people who are overweight are hungry all the time, but some of them don't notice when they are feeling hungry. That means when they do eventually eat, they are so hungry that they lose control and will over eat. Other people may think they feel hungry when, in fact, they feel thirsty. In that case, the best thing to do is drink

The danger of diets

Dieting creates feelings of deprivation making people want a particular food even more than they would normally.

Diets come with rules. Our Reticular Activating System (RAS) monitors our goals and progress – including negative ones like not, say, eating chocolate. So we check frequently, resulting in us thinking about chocolate more than we normally would.

Diets make us stressed about food and eating. And stress fires the HPA (Hypothalamus-Pituitary-Adrenal) axis – putting us in fight or flight mode.

And if we're not eating as much as we used to, we must be in a starvation situation. The best way to cope with that is to eat as much as we can – and that's what happens when we stop dieting and stop having so much control over what we eat.

a glass of water and wait to see whether the pangs of hunger go away. Some people eat when they are tired. And that's why helping clients to get a good night's sleep is so important. Some people eat because of emotional issues. The miracle question can help find out more about what's going on in their life that they find so upsetting. And some people eat for secondary gain – being fat and less able or attractive brings some other advantage you might not at first guess.

So, why do people eat? It seems there are a number of reasons:

- Because they are hungry and need nutrients and energy
- Because they like the taste of something and so eat more because they enjoy it
- Because they are bored, and eating gives them something to do
- Because they use food for emotional support (comfort eating)
- Because other people are eating
- Because food is available
- Because it's a special occasion
- Because they're tired
- Because it's meal time.

Let's have a look at the second of those reasons to eat, because you like the taste of something (like chocolate). What's going on in your brain when your hedonic appetite system is working – when your dopamine reward system is in play. It seems that eating is a bit like "chasing the dragon". We need to keep eating more and more of something we like to get the same amount of pleasure as the first time we ate it That's what a study (by Stice *et al* 2010) found. This was confirmed by Wilcox *et al* (2010). They found that high BMI and difficulty controlling weight are associated with low dopamine activity.

Pepino *et al* (2016) found that in general, people grow less fond of sweet things as they move from adolescence into adulthood. Also, as people age, they have fewer dopamine receptors in the striatum, that is critical to the reward system. The study found that both younger age and fewer dopamine receptors are associated with a higher preference for sweets in those of normal weight, but not in people with obesity.

As therapists, we need to ensure that clients identify other rewarding activities in their life in addition to food. This will help them to produce and maintain dopamine brain activity and takes the focus off food.

The body is designed to use fat and carbohydrate in creating energy. So eating fat isn't bad. Eating more fat than can be used results in excess fat being stored. So, what many people do is cut out fat and eat excess carbohydrates. The problem is that excess carbohydrates are also stored as fat. So the secret of healthy eating is to eat both carbohydrates and fats, but not too much of them. Fat isn't bad – after all, many vitamins (A, D, E, and K) are fat soluble.

You may well see people with eating disorders. The most common ones you'll come across are:

- Anorexia nervosa – when a person tries to keep their weight as low as possible; for example by starving themselves or exercising excessively.

- Bulimia – when a person goes through periods of binge eating and is then deliberately sick or uses laxatives (medication to help empty the bowels) to try to control their weight.

- Binge Eating Disorder (BED) – when a person feels compelled to overeat large amounts of food in a short space of time.

- Orthorexia nervosa – a proposed eating disorder characterized by an excessive preoccupation with eating 'healthy' food. Some people's dietary restrictions intended to promote health may paradoxically lead to unhealthy consequences, such as social isolation, anxiety, loss of ability to eat in a natural and intuitive manner, reduced interest in the full range of other healthy activities, and, in rare cases, severe malnutrition or even death.

- EDNOS – an Eating Disorder Not Otherwise Specified (EDNOS). This means people have some, but not all, of the typical signs of eating disorders like anorexia or bulimia.

You may also see people with Body Dysmorphic Disorder (BDD) – an anxiety disorder that causes a person to have a distorted view of how they look and to spend a lot of time worrying about their appearance. For example, they may be convinced that a barely visible scar is a major flaw. This may impact on what they eat.

Eating disorders can result from using food to cope with painful emotions such as anger, self-loathing, vulnerability, and fear. They can be a coping mechanism – refusing food might make people feel in control, binge eating might be comforting, or purging might be a way of punishing themself. What's needed are healthier ways to cope with negative emotions. Once a client has identified the emotion they are experiencing, they can choose alternative ways of coping with that emotion.

So, what should you do when treating your clients? What's the best way to get the best results for them? The first thing to say is to avoid being seduced by the problem. Just because they say they want to lose weight, don't spend the whole time talking about food and weight loss.

I do weigh people when they come to see me. I do have some weight-loss scripts, which I use for the first couple of sessions. The reason for this is because it proves to them that I am listening to them and their issues. It helps to build rapport. It's also pretty much what they would expect to happen. And it motivates them with something they want to move 'away from'.

Once we have established trust and a therapeutic alliance, we can focus on their goals. And, quite often, their real goal may not be revealed until session 4 or later.

So, that's the theory, what can you actually do with your clients? One powerful question to ask clients is: "What stands in the way of (your) success?" It's a great way of getting them thinking and it helps shape the work that you need to do with them. But, as I said earlier, the important thing to remember is that you should, in the main, apply exactly the same principles as you would when dealing with any other problem.

Why is your client overeating? What triggers their eating problem? What beliefs do they have about food and eating that need to be changed? What limiting beliefs do they have about their ability to lose weight that need to be overcome? What's their self-esteem like? What habits do they have that need to be changed? Do they recognize and acknowledge that they have an eating disorder (if they do have one)? How well are they sleeping?

Our standard approach is to help people reduce their levels of anxiety. This can make them better able to control their impulses (eg snacking). And we can help them to sleep better. The *Journal of Clinical Endocrinology & Metabolism* (18 Jan 2012) published an article showing a direct link between sleep deprivation and obesity. While it's good to be solution focused and not spend too much time

Exception questions

Has anything been better since the last appointment? What's changed? What's better?

Can you think of a time in the past (month/year/ever) that you did not have this problem?

What would have to happen for that to occur more often?

When doesn't the problem happen?

What's different about those times?

What are you doing or thinking differently during those better times?

When have you been able to stop doing....?

Are there times when you expect to...but you remember something that helps you calm down?

What else?

talking about food, talking about exercise may not be successful with people who were never very sporty at school and have always tried to avoid it

For many people, simply eating a little less will help cut down the calories they take in each day. If they decrease the amount of food they eat dramatically, this could have health implications. It can also lead them to feeling hungry and then uncontrollably binge eating and reducing their self-esteem because they feel they don't have any control.

As well as the usual sparkling moment, scaling, and miracle question, it's useful to ask exception questions and coping questions. The exception questions can identify when the client doesn't have the eating disorder or doesn't eat too much. The coping questions identify how the client successfully copes with the eating disorder.

So, what should be the focus of our goal-oriented work? The answer will be determined by the clients' answers to the miracle question, but I would suggest that being healthy is something that everyone can aspire to.

During the trance sessions, you should try to include ideas and phrases from the client to personalize any scripts (language patterns) you're using. And you can spend time helping the client visualize (rehearse) choosing healthy foods and healthy eating techniques. You can also help them visualize doing healthy activities such a going for a walk or, if they are up to it, going swimming or going to the gym and enjoying the activity.

You may need to spend time helping them to change bad habits and replace them with better ones. The trick here is to get them to visualize each stage of the new habit rather than just visualize the successful conclusion of the change.

It's also useful to encourage them to spend time with thinner people who eat well and exercise, and not spend time with larger people who eat less healthily and probably don't consider exercise.

One suggestion is to use metaphors in which outer layers are purposefully discarded. You might like to talk about sculptors chipping away at stone, wood, ivory, or bone, and finding something beautiful inside that the sculpting process will reveal. Or, perhaps, they're discarding layers of heavy winter clothes to reveal their beach-ready body below (or at least their cooler summer clothing). You might send them on a journey carrying a heavy rucksack and numerous other bags they can gradually discard as they travel on that journey. They struggle with the

Coping questions

How do you cope with these difficulties?

What keeps you going?

Who is your greatest support?

What do they do that is helpful?

What do you do that stops the problem getting worse?

When you've had this problem before, what helped you get through then?

How did you manage to solve the problem?

What advice would you give to someone else who has this problem?

What else?

weight to begin with, but as they discard more things, they feel lighter and their whole body feels healthier.

It's also useful to get them to imagine a future version of themselves, who has achieved their goals, and get that future self to tell them what steps they took to achieve this desired state. That helps to give them a realistic idea of what they have to do in terms of looking good, eating well, and exercising.

It can also be useful to revisit how they feel when they are about to eat a takeaway or a big meal. Are they thinking about the taste of the food or its texture? Or are they thinking about how the meal will feel in their stomach – a sort of heaviness that will last for a while after the meal has finished. Eating a big meal is like falling out of tree. Everything seems fine until the last moment (when you hit the ground). So, clients should consider the consequences of eating something before they start.

It's useful to empower your client. Diets tend to be restrictive about what people eat. You're not going to say what they can and can't eat. You'll encourage them to make sensible decisions about what they eat and what they don't eat. But if they do want to eat something like a takeaway, they can – just once in a while.

But be prepared to find more going on in their life that they may want to discuss, in addition to eating too much.

References:

http://www.uncommonhelp.me/articles/why-hypnotherapy-works-for-weight-loss/

http://www.uncommonhelp.me/articles/weight-loss-motivation/

https://medicine.wustl.edu/news/age-obesity-dopamine-appear-to-influence-preference-for-sweet-foods/

http://www.sparkpeople.com/resource/nutrition_articles.asp?id=1660

Biome brain

A look at the bacteria that live in your GI tract and how you affect them and how they affect you. These bacteria can affect people trying to lose weight and people with low mood.

Your gut biome is the naturally occurring bacteria that live inside your gastrointestinal tract. Your gut, that long tube that goes from your oesophagus to your anus is a fascinating tube in its own right because it contains as many neurons lining it as there are in a cat's brain – over 100 million. The microbes' job is to orchestrate digestion and moderate gut pain. You have two kilos of microbes living in your gut. That's around 50 trillion individual bacteria from around a thousand different species.

Your gut has to let food through it, but mustn't let anything else through. And it needs to maintain a heathy barrier – if it becomes leakey, then you become ill. The bacteria living in your gut protect your gut from invaders and regulate your immune system. Changing the mix of bacteria in your gut can reduce the number of coughs and colds you get. These gut bacteria also help to regulate your body weight because they control how much energy you can extract from food, they affect how hungry you feel, and they can determine which foods you crave. In addition, they can affect how big a blood sugar spike you get after eating. And, lastly, they can convert indigestible food into hormones and other chemicals.

It's interesting to see how different foods are digested. For example, sugary drinks pass quickly through the stomach into the intestine, where the sugar is extracted and absorbed. This can lead to a very quick sugar spike in the blood. But if you have sweetened drinks (using an artificial sweetener) rather than sugary ones, they can cause inflammation in the gut. Popular carbohydrates like bread, pasta, potatoes, and rice, because they are low in fibre, also spend only a short time in the stomach before they pass into the intestines and are digested into simple sugars that are absorbed – causing a blood sugar spike. Protein, fat, and fibre spend longer in the stomach and take longer to digest.

Alcohol can irritate the stomach lining, which makes the blood vessels swell. This increases the surface area and increases the speed and the amount of alcohol that's absorbed. Once it's absorbed, the liver starts to break down the alcohol, first into acetaldehyde, which causes red faces and hangovers, then acetic acid, and finally CO_2 and water.

Phytochemicals are chemical compounds produced by plants and are classified into major categories, such as carotenoids and polyphenols, which include phenolic acids, flavonoids, and stilbenes/lignans. Flavonoids can be further divided into anthocyanins, flavones, flavanones, and isoflavones, and flavanols.

Carotenoids and flavonoids can be found in yellow, orange, red, blue, and purple fruits and vegetables. Polyphenols are found in cocoa, olives, dark chocolate, tea, coffee, and red wine. Flavonoids are antioxidants and they encourage your body to burn fat.

The hormone ghrelin makes you feel hungry, and leptin makes you feel full. Leptin is made by adipose cells, so when you cut out fat from your diet, you make less of it – and so feel hungrier. Also, when you diet, you produce more ghrelin (making you feel hungrier). PYY is another hormone that suppresses appetite and slows down the emptying of the contents of the stomach into the intestines. The amount of PYY you make can be increased by eating more protein and by eating more slowly.

Prebiotics are foods that feed good bacteria

Probiotics are foods containing good bacteria

Let's take a more detailed look at your biome – those bacteria that live in your gut. Most people have a mixture of good bacteria and which ones predominate is determined by what they eat and how varied their diet is. Good bacteria include:

* Firmicutes – these help digest fats, and their predominance in your gut can be linked to obesity

* Bacteroidetes – these are associated with leanness when plentiful. They teach your immune system how to behave, and break down undigested fibre to produce butyrate. Butyrate controls the growth of the cells in gut walls. It also has an anti-inflammatory effect. Their numbers can be increased by eating lots of fibre.

* Akkermansia – these live on the mucus secreted by the gut wall. They strengthen the gut wall and reduces inflammation. Their numbers can be increased by eating polyphenol-rich food. They also thrive when you eat less food.

* Christensenella – these are found in people who tend to be lean.

* Lactobacillus – these protect your guts from fungal pathogens.

* Bifidobacterium – these help to break down indigestible fibre, and protect your gut from other microbes. They are found in cheese and yoghurt.

The appendix acts as a reservoir of good bacteria.

The GI tract also hosts unfriendly bacteria, the ones that make you ill. These include:

Shannon's Index and Simpson's Index show how diverse your biome is.

* Campylobacter – these cause food poisoning

* E coli – these can cause bloody diarrhoea

* Salmonella – these cause food poisoning

* Clostridium difficile – these cause watery diarrhoea, fever, and abdominal pain. They often predominate in the gut after a course of antibiotics.

A person with a diverse microbiome will be heathier because it's in the interest of the gut bacteria to keep the host healthy. To ensure a person has a diverse biome, they should eat as many different plant varieties as possible. Changing your diet can change the mix of bacteria in your gut. Taking antibiotics can reduce the number of gut bacteria. Eating processed food can also impact on your biome because the emulsifiers used to extend the food's shelf life are bad for your biome. What's good for your bacteria is to open the window, and do gardening.

Inulin, which can be found in onions leeks, garlic, and chicory, is a good prebiotic and promotes a healthier biome. It also reduces constipation, and is good for bones.

Sources of probiotics include yoghurt and cheese, fermented foods like kefir and sauerkraut, and apple cider vinegar.

What many people don't appreciate is that two different people can eat the same food, but it can have a different effect on their blood sugar levels. And it's all due to their different biomes. Some gut bacteria are able to extract more energy from food than others. That's what causes different blood sugar levels, and that can affect your mood. In addition, your biome affects which foods you like.

A sugar crash is the feeling of fatigue or lethargy after consuming large amounts of carbohydrates. Firstly there's a rapid rise in blood glucose after eating, which leads to insulin being secreted (the insulin spike), which results in the glucose being removed from the blood (by the liver or it being used by the cells). This creates the sugar crash. One consequence of a sugar crash is that people eat more (to compensate). One way to prevent this is to cut down on sugary drinks and snacks and anything with a high Glycaemic Index (GI).

Psychobiotics are gut bacteria that communicate with the brain through the enteric nervous system, the vagus nerve, the immune system, and hormones in the gut.

Gut bacteria produce the neurotransmitters dopamine, serotonin, and GABA. There is plenty of evidence to suggest that these microbes can manipulate our behaviour and mood by altering the neural signals in the vagus nerve. They can change a person's taste receptors, they can produce toxins to make us feel bad, and they can release chemical rewards to make us feel good. In addition, microbes control what we eat and how much of it we eat.

Researchers have found that eating leads to widespread opioid release in the brain, likely signalling feelings of satiety and pleasure.

Growing evidence suggests that, at least, some forms of depression may also be linked to inflammation in the body. Therefore, it makes sense to eat foods that are anti-inflammatory (like tomatoes, spinach, salmon, etc).

Sleep is important for the intestinal lining to repair itself and to create a healthy microbiome. Lack of sleep makes you hungry and leads to stress, which leads to sleep deprivation.

It makes sense, when helping clients with weight loss and understanding why they feel the way they do, that we explain about the bacteria living in their gut and how those bacteria are impacting on what they eat, the effect food has on their body, and how that is affecting their mood.

References:

The Gut Microbiome, Anxiety and Depression: 6 Steps to Take: https://www.
psychologytoday.com/blog/inner-source/201411/the-gut-microbiome-anxiety-and-
depression-6-steps-take

https://www.psychologytoday.com/blog/urban-survival/201701/new-research-shows-
depression-linked-inflammation

https://www.utu.fi/en/news/news/Pages/Eating-Triggers-Endorphin-Release-in-the-
Brain.aspx

Dr Michael Mosley. The Clever Guts Diet: How to revolutionise your body from the
inside out. 978-1780723044

Superfoods

Here we take a look at superfoods and whether they can help our clients. Trevor Eddolls has a diploma in nutrition.

We see clients for weight management (a euphemism for losing weight) and we may find it useful to have a conversation with them about what they eat (in addition to helping them empty their buckets, relax, and get control of their lives). And we know that, for many of our clients, when they get that triumvirate of sleeping, exercise, and eating right, they are better able to get back control of their lives. So, for them, talking about when and what they eat may also be useful. So wouldn't it be great if we could discuss superfoods with them – especially if that would somehow help them.

So, what is a superfood? Well, basically, it's described as a nutrient-rich food that's considered to be especially beneficial for health and wellbeing. What there isn't is an official definition as such, a hurdle that a foodstuff has to get over in order to become an official superfood. And that means you find all sorts of foods being labelled as superfoods by all sorts of people. Beer and chips could be called superfoods!

> Antioxidants inhibit the oxidation of other molecules. Oxidation can produce free radicals, which lead to further reactions that may damage cells.

However, it's worth bearing in mind that since 2007, nothing can be marketed as a 'superfood' unless there is an accompanying specific authorized health claim supported by credible scientific research. This was an EU ruling.

> Most polyphenols are good for the body (eg reducing inflammation) but some are considered antinutrients, compounds that interfere with the absorption of essential nutrients, especially iron and other metal ions.

If you look on the Internet, you'll find there are some foods that regularly appear on lists of superfoods. And, as I said before, there is no single list of what foods are superfoods. What they tend to have in common is that each is described as some kind of nutrient powerhouses that punches above its weight in terms of the amounts of antioxidants, polyphenols, vitamins, or minerals that it contains. And people who eat these superfoods do so in the expectation that they are reducing their risk of chronic disease, and that they will be thinner, healthier, and live longer than people who don't.

So lets have a look at the sorts of foods that people are claiming to be superfoods:

- Almonds – contain fibre, potassium, calcium, vitamin E, magnesium, and iron.
- Apples – are full of fibre.
- Beans – are high in protein, fibre, folate, and magnesium. Studies indicate that legumes can lower cholesterol and reduce the risk of certain cancers.
- Beetroot – is full of vitamins, minerals, and antioxidants that can help fight disease and strengthen vital organs. Betalains (the purple pigments) may help ward off

cancer and other degenerative diseases. Beetroot, for example, is a good source of iron and folate (naturally occurring folic acid). It also contains nitrates, betaine, magnesium and other antioxidants (notably betacyanin). Beetroot may help lower blood pressure, boost exercise performance and prevent dementia.

- Blueberries – are filled with fibre, vitamin C, vitamin K, manganese, and antioxidants (notably anthocyanins). They fight against heart disease, some cancers, and may even improve memory.

- Broccoli – is packed with vitamins, minerals, disease-fighting compounds, and fibre. It has high levels of vitamin C and folate (which can reduce the risk of heart disease, certain cancers, and stroke).

- Cauliflower – contains vitamins and minerals, and glucosinolates. These are phytochemicals that can prevent damage to the lungs and stomach by carcinogens, potentially protecting against those cancers. And may also help prevent hormone-driven cancers like breast, uterine, and cervical.

- Chia seeds – have the most essential fatty acids of any known plant, plus they come with magnesium, iron, calcium, and potassium.

- Cocoa/chocolate – is a good source of iron, magnesium, manganese, phosphorous, and zinc. Cocoa also contains the antioxidants catechins and procyanidins. It may lower blood pressure, and protect against cancer and stress.

- Cranberries – can help fight inflammation, reduce the risk of heart disease, improve oral health, help prevent ulcers and yeast infections, and may even inhibit the growth of some human cancer cells.

- Eggs – are full of protein and omega-3 fatty acids.

- Garlic – contains vitamins C and B6, manganese, selenium, and other antioxidants (notably allicin). It can be used to treat anything from high blood pressure and heart disease to certain types of cancer. Plus, studies suggest garlic extract can be used to treat yeast infections in women and prostate issues in men.

- Greek yogurt – is full of protein and probiotics. It improves digestion, and is good for the immune system.

- Green tea – is full of antioxidants (including catechins) and contains B vitamins, folate (naturally occurring folic acid), manganese, potassium, magnesium, and caffeine. Epigallocatechin gallate (EGCG) is a phytochemical that slows irregular cell growth, which could potentially help prevent the growth of some cancers. It may also boost weight loss, reduce cholesterol, combat cardiovascular disease, and prevent Alzheimer's disease.

- Ginger – a remedy for everything from an upset stomach to unwanted inflammation.

- Goji berries – contain vitamin C, vitamin B2, vitamin A, iron, selenium, and other antioxidants (notably polysaccharides). They may boost the immune system and brain activity, protect against heart disease and cancer, and improve life expectancy.

- Kale – provides more antioxidants than most other fruits and vegetables, and is a great source of fibre, calcium, and iron.

- Leeks – contain organosulphur compounds, which may fight against cancer and boost immunity.
- Lentils – are high in protein, iron, and other essential nutrients.
- Oatmeal – is high in fibre, antioxidants, and other nutrients. It can lower cholesterol levels, aid in digestion, and improve metabolism.
- Pistachios – are full of protein, fibre, and potassium.
- Pomegranates – are full of fibre, and contains vitamins A, C, and E, iron, and antioxidants (notably tannins). They may be effective against heart disease, high blood pressure, inflammation, and some cancers, including prostate cancer.
- Pumpkins – are full of antioxidants and vitamins. They also contains beta-carotene, a provitamin that gets converted to vitamin A.
- Quinoa – provides all nine essential amino acids.
- Salmon – is full of protein and omega-3 fatty acids, which reduce the risk of cardiovascular disease, prostate cancer, age-related vision loss, and dementia.
- Spinach – contains antioxidants, anti-inflammatories, and vitamins.
- Strawberries – are full of vitamin C. Its antioxidant helps build and repair body tissues, boost immunity, and fight excess free radical damage.
- Watermelon – is high in vitamins A and C. It may lower blood pressure and reduce the risk of cardiovascular disease. The lycopene may protect the body from UV rays and cancer.
- Wheatgrass – contains chlorophyll, vitamin A, vitamin C, vitamin E, iron, calcium, and magnesium. It may protects against inflammation, build red blood cells, and improve circulation.

That all sounds brilliant. But what's the evidence?

Blueberries

Interestingly, a 2012 study of 93,000 women found that participants who ate three or more portions of blueberries and strawberries a week had a 32% lower risk of a heart attack compared with those who ate berries once a month or less. Unfortunately, the study could not prove that these fruits definitely caused the lower risk.

There's also inconclusive evidence that blueberries may relax the walls of the blood vessels, which may help reduce this risk of atherosclerosis (hardening of the arteries), which may, in turn, reduce the risk of a heart attack and stroke.

A 2015 study of 48 post-menopausal women, found that women who were given blueberry powder supplements over the course of eight weeks experienced a small, but clinically significant, drop in blood pressure. However, there was another study that year of 44 adults with metabolic syndrome (a combination of diabetes, high blood pressure, and obesity). They were given blueberry smoothies, which had no effect on blood pressure.

In laboratory studies on cells and animals, blueberry extracts (such as anthocyanins) have been shown to decrease free radical damage that can cause cancer. However, it is not clear how well humans absorb these compounds from eating blueberries and whether or not they have a protective effect.

A number of small studies have found a link between blueberry consumption and improved spatial learning and memory. However, most of these studies relied on small sample groups or animals. There is currently no evidence of a link between eating blueberries and improved memory.

Beetroot

A 2010 small and short-term study suggested that a diet high in beetroot juice may increase blood flow to certain areas of the brain.

A 2013 review of the existing evidence concluded that beetroot juice was associated with a modest reduction in blood pressure. Beetroot is rich in nitrates. When ingested, our body converts nitrates into nitric oxide, which lowers blood pressure.

Also in 2013, a review found that inactive and recreationally-active individuals saw "moderate improvements" in exercise performance from drinking beetroot juice, although it had little effect on elite athletes.

A 2014 study investigated the effects of beetroot juice on cyclists, who were cycling in a chamber designed to mimic the effects of relatively high altitude (2,500 meters above sea level). The juice had a modest but significant increase in terms of the cyclists time trial scores – on average there was a 16 second improvement.

Goji berries

There is no reliable evidence to suggest goji berries improve immunity, cardiovascular disease, and life expectancy. Most of the research used small-sized trial groups, were poor in quality, and were performed in laboratories using purified and highly concentrated extracts of the goji berry.

One 2008 study of 34 people found a daily drink of 120ml of goji berry juice for 14 days improved feelings of wellbeing, brain activity, and digestion.

A 1994 Chinese study on 79 patients with various advanced cancers found those treated with immunotherapy in combination with goji polysaccharides saw their cancers regress. Unfortunately, there's no information on the design of the study and the goji berry compounds used.

Cocoa/chocolate

A 2012 review of the effects of chocolate on blood pressure concluded that cocoa products, including dark chocolate, may help slightly lower blood pressure.

Some limited animal and laboratory research suggests a cocoa-rich diet could offer protection against bowel cancer.

A 2009 study of 30 healthy people given 40g of dark chocolate a day for 14 days experienced a reduction in stress hormones.

A 2015 study reported that people who ate the equivalent of two chocolate bars a day had a slightly lower risk of stroke than people who never or rarely ate chocolate.

Garlic

A 2012 review identified one study suggesting that 200mg of garlic powder three times daily reduced blood pressure. However, the review concluded there was insufficient evidence to say whether garlic was an effective means for treating high blood pressure and reducing death rates.

A 2009 review concluded that garlic – mainly garlic powder – produced "modest reductions" in total cholesterol levels.

A 2012 review concluded there was insufficient evidence regarding the effects of garlic supplements in treating or preventing colds.

A 2007 World Cancer Research Fund review concluded that garlic "probably protects against" bowel and stomach cancers. A 2009 review concluded there was "no credible evidence" with stomach, breast, lung, and womb cancers, but that there was "very limited evidence" that eating garlic may lower the risk of colon, prostate, oral, ovary, or renal cell cancers.

Green tea

A 2009 review found that there's no evidence drinking green tea protects against different types of cancer. A 2015 study looked at the cancer-fighting effects of a compound found in green tea when combined with Herceptin (used to treat stomach and breast cancer). Initial results in the laboratory were promising.

It's thought the antioxidants catechin and caffeine found in green tea may have a role in helping the body burn more calories (speeds up the metabolism), which may help weight loss. However, a 2012 review found no significant effect of weight loss from drinking green tea.

A 2010 animal study found a green tea preparation rich in antioxidants protected against the nerve cell death associated with dementia and Alzheimer's disease.

Another 2011 review found drinking green tea enriched with catechins led to a small reduction in cholesterol.

A 2013 review found daily consumption of green and black tea could help lower cholesterol and blood pressure thanks to its catechins.

A 2014 review found evidence of a modest reduction in blood pressure in people with high blood pressure who consumed green tea.

A 2014 study looked at how effective a green tea mouthwash was in preventing tooth decay compared with the more commonly used antibacterial mouthwash chlorhexidine. The results suggested they were equally effective.

Oily fish such as salmon, mackerel, and sardines

A review in 2004 by the UK Scientific Advisory Committee on Nutrition concluded that a "large body of evidence" suggests that fish consumption, particularly oily fish, reduces the risk of cardiovascular disease.

Studies have found eating oily fish can lower blood pressure and reduce fat build-up in the arteries. The Government recommends that people eat at least two portions of fish a week, one of which should be oily.

The evidence for oily fish's effect on prostate cancer is inconclusive.

A 2010 review found some evidence that eating oily fish two or more times a week could reduce the risk of age-related macular degeneration – a common cause of blindness in older people. A 2015 review looked at whether fish oil supplements could reduce the progression of macular degeneration in people who already had the condition found no evidence of any benefit.

A 2012 review concluded that there is no preventative effect or decline in brain function and dementia when healthy older people take omega-3.

A 2013 study found that women who ate one or more servings of oily fish were 29% less likely to develop rheumatoid arthritis than women who never, or very rarely, ate oily fish.

A 2013 review by NICE (National Institute for Health & Care Excellence) was inconclusive whether omega-3 fatty acids could improve the symptoms of schizophrenia.

Pomegranates

A 2006 study found that drinking a daily 227ml glass of pomegranate juice significantly slowed the progress of prostate cancer in men with recurring prostate cancer.

A 2013 study on mice found evidence that pomegranate strengthened bones and helped prevent osteoporosis.

Another 2013 study looked at whether giving men pomegranate extract tablets prior to surgery to remove cancerous tissue from the prostate would reduce the amount of tissue that needed to be removed. The results were not statistically significant.

A 2004 study on patients with carotid artery stenosis (narrowed arteries) found that a daily 50ml glass of pomegranate juice over three years reduced the damage caused by cholesterol in the artery by almost half, and also cut cholesterol build-up.

A 2005 trial on patients with coronary heart disease demonstrated that a daily 238ml glass of pomegranate juice administered over three months resulted in improved blood flow to the heart and a lower risk of heart attack.

Wheatgrass

There is no evidence to show taking wheatgrass juice enhances haemoglobin production.

A 2002 study found patients with ulcerative colitis (inflammation of the colon) saw their symptoms improve after they were given 100ml of wheatgrass juice daily for a month.

A 2004 study of patients with a blood disorder called thalassaemia found half of the patients required fewer blood transfusions when 100ml of wheatgrass juice was taken daily for three years.

Diet and mood

Improving your diet helps improve your mood. Here are some things to look for:

- Omega-3 fats are found in fatty fish, omega-3 fortified eggs, and supplements. The omega-3 fat DHA is important for neuron structure. The omega-3 fat EPA helps with neuron function and also reduces inflammation.
- Iodine is critical for a healthy thyroid. Good food sources include seaweed, cod, and iodized salt. There's also some in milk, yogurt, and eggs.
- Zinc is involved in over 250 separate biochemical pathways in the body. It is critical for neurotransmitter production and function. Iodine is found in oysters, crab, beef, lamb, pork, dark meat, chicken, legumes, and cashews.
- Magnesium is required for over 300 separate biochemical pathways. Magnesium helps activate the enzymes needed for serotonin, dopamine, and noradrenalin production. Zinc is found in nuts and seeds, dark green vegetables, whole grains, bran, and dark chocolate.
- Vitamin D deficiency has not only been linked to depression but anxiety, SAD, and dementia as well. Vitamin D is found in oil/fatty fish like salmon, sardines, mackerel, herring, and trout, and eggs to a lesser extent.
- Selenium is needed for good thyroid function. It also helps make glutathione. Increasing glutathione levels reduces depression, probably by reducing inflammation in the brain. Selenium is found in Brazil nuts, fish, ham, prawns, liver, and chicken.

- Iron deficiency can lead to anaemia, which can lead to depression, fatigue, irritability, apathy, brain fog, and lack of motivation and appetite. Iron can be found in beef, pork, lamb, dark meat chicken, eggs, liver, oysters, and white beans.

- B Complex (there are around 11 B vitamins) is involved in neurotransmitter production and function. B12 maintains brain mass and prevents brain shrinkage. A lack of B12 can lead to dementia and depression. Folate, B12, and B6, help lower the levels of homocysteine, a by-product of protein metabolism. High levels of homocysteine increase the risk of depression. B vitamins can be found in whole grains, nuts and seeds, dark green vegetables, and meat.

- Vitamin C deficiency not only leads to scurvy and bleeding, swollen, and achy gums, but also depression. Vitamin C can be found in citrus fruits, kiwi, bell peppers, and strawberries.

References:

http://www.nhs.uk/Livewell/superfoods/Pages/superfoods.aspx

http://greatist.com/health/25-greatist-superfoods-and-why-theyre-super

https://www.curejoy.com/content/nutrient-deficiencies-cause-depression-05-2017/

You wouldn't like me when I'm angry

Here are some ideas to help clients who come with issues around anger.

Anger is a very natural feeling. It's one of the ways that the primitive brain deals with continued (chronic) stress. It provides animals and people with an evolutionary advantage. When they're angry, they are stronger and are better able to defend their food and family – and so it is more likely that their genes will be passed on to the next generation. And appearing angry can act as a warning to others to modify their behaviour

It's important to separate the feeling of being angry from any behaviour. While for our ancestors, anger would have given them an evolutionary advantage, bouts of anger today are usually inappropriate and can result in social exclusion.

Let's recap how the brain works. There are three parts to the brain. The brain stem controls all those useful functions such as keeping your heart beating and ensuring you breath. Your primitive brain (pretty much the limbic system) is a fast-acting part of your brain and ruled by emotions such as fear, disgust, happiness, sadness, surprise, and anger. And the third part (your cerebral cortex) is the slower intellectual brain, which is able to make logical decisions. When you're stressed, you're less able to use your intellectual brain and, eventually your primitive brain protects you with depression, anxiety, or anger.

It seems that during an angry episode, the left hemisphere is strongly activated, but not much happens in the right hemisphere. This gives you some logical ability, but no contextual ability. And that's why people do things that seem sensible to them at the time, but which later they come to bitterly regret.

Extreme anger can damage the heart and the immune system, whether it's released or repressed (even remembering times you felt very angry can be bad for your heart). Constantly releasing anger isn't good because the more you do something, the more likely you are to do it. So, you become more likely to respond to any situation by becoming angry.

A client seeing you with anger issues will find that they behave in a way that is unhelpful and destructive. And the anger is impacting on their overall mental and physical health. Your client may be outwardly aggressive, inwardly aggressive, or even passively aggressive.

Apart from on-going stress, people may get angry because they feel threatened or are attacked. They may be feeling frustrated or powerless. Or they may feel as if they're being treated unfairly or slighted in some way. They may fear that they have been abandoned or they may feel overwhelmed. Remember it's not events themselves that affect us, it's how we interpret them. And so people may treat events as being more threatening or more unfair that they might actually be.

What can we, as hypnotherapists, do to help our clients? The obvious things are to help

> Passive aggressive behaviour is generally non-verbal negative behaviour, eg sullenness, stubborness, or failure to complete tasks.

them empty their stress buckets and get them into their intellectual brains. That means getting them to exercise, eat properly, and get enough sleep. It also means helping them to relax – and encouraging them to notice good things, not just bad ones. And that's what we would do in a standard session. But is there anything else we could do?

The obvious one is to get them to recognize triggers. Ask them what kinds of situation make them angry. How did they feel at the time; how did they behave; and how did they feel afterwards? And discuss how might those triggers be interpreted in a different way. For example, if someone in their office made tea for everyone else, the client might think they were ignored because no-one liked them. An alternative explanation might be that they didn't look up when the person asked if anyone else wanted a drink. It may be that they still had some tea in their cup and the person making the drink assumed they had one. Once a client understands that alternative explanations are possible, they are less likely to become so angry. Also, encourage your client to stop using words like 'always', 'never', 'must', 'ought', and 'should', and replace them with the phrase, "I would prefer it if...".

The second thing is to get them to recognize the signs of becoming angry and take steps before they go into full-blown anger outburst. Signs include: faster heartbeat, faster breathing, feeling tense, clenched jaw or fists. This then gives them a chance to walk away from the situation or count to 10 before the anger takes over. It gives their intellectual brain a chance to think about how they want to respond.

Two ways to avoid getting angry are gratitude and forgiveness. If you regularly think of all the things you are grateful for, the usual triggers are less likely to make you cross. Similarly, if you forgive the people who make you angry, you've removed the emotional baggage you usually associate with them, so you will be much less likely to be angered by them.

Thirdly, you can get them to imagine other responses they could have made in a situation that had previously angered them. Then get them to rehearse (future pace) how they would like to behave in situations in the future. You can get them to visualize what it might be like to be the other person that they are venting their feelings on. Imagine that the other person has a life, worries, feelings, problems, and fears.

There are other techniques that you could teach your client that can help them to calm down. These include:

- Breathing techniques such as 7-11 breathing where they breathe out for longer than they breathe in and they focus on each breath they take.
- Relaxation techniques such as tensing each part of the body in turn and then relaxing it.
- Going for a run, a swim, or a brisk walk to burn off the adrenalin in their body
- Distracting themself by dancing to upbeat music, colouring in, or taking a cold shower.
- Using Mindfulness techniques to recognize when they're getting angry and to calm down.
- Using a mantra such as 'relax' or 'keep calm'.

- Trying Yoga.
- Visualizing calm walks on the beach and other places where they've felt calm and in control.
- Reading jokes and laughing.
- Making a physical change, eg walking over to a window where the sun is shining.

You'll find that some clients resort to violence because their communication skills are weak or because their Emotional Intelligence skills aren't too good. They don't know how to be assertive and respectful and so situations can get out of hand. Some techniques you might look at include helping your client to:

- Think about the outcome they want to achieve.
- Use phrases starting with "I feel…", eg "I feel angry with you because…". That avoids the other person feeling that you are blaming them.
- Listen to the other person's response and try to understand their point of view.
- Prepare for the conversation to go wrong and plan how to stop the conversation before getting angry.

You might want to encourage your client to avoid drugs and alcohol – if that's an issue for them.

What can you do if your client gets angry during a session? There are some techniques that can be used, including:

- Staying calm and in control. Don't mirror their anger. Recognize that it's their preferred strategy.
- Identifying why they are angry
- Validating their anger. Apologize for their bad experience.
- Helping them to express in words just how angry they are.

If you're working with children, you need to separate the anger from the child. You could give their anger a name for example 'volcano'. Agree on a goal that you're both working towards. Give the child positive feedback when they are in control of their anger. Help them to recognize emotions in others. Don't dismiss feelings, they can't help how they feel. What they can control is how they act. So acknowledge the feeling and help them to achieve calmness. Ask the child to suggest other ways that they could have behaved instead of being angry in a situation. Help the child to empathize with how other people are feeling. You might need to suggest alternative strategies as part of a conversation. Show the child some relaxation techniques.

If you have a conversation with the parents, it can be worthwhile suggesting that they move to a zero-tolerance policy and don't tolerate the aggression. This does not mean that they should be aggressive themselves and smack the child, just that the child's actions should have consequences, like losing TV or tablet time. This will help them to gain control of their behaviour, while still allowing them to have feelings. They will learn that feelings aren't bad, just some behaviours are less successful.

References:

http://www.nhs.uk/Conditions/stress-anxiety-depression/Pages/controlling-anger.aspx

http://www.mind.org.uk/information-support/types-of-mental-health-problems/anger/

https://www.psychologytoday.com/blog/happiness-in-world/201311/dealing-anger

http://www.apa.org/topics/anger/control.aspx

http://www.mayoclinic.org/healthy-lifestyle/adult-health/in-depth/anger-management/art-20045434

http://www.nhs.uk/Conditions/stress-anxiety-depression/Pages/dealing-with-angry-child.aspx

http://tinybuddha.com/blog/20-things-to-do-when-youre-feeling-angry-with-someone/

https://www.supernanny.co.uk/Advice/-/Parenting-Skills/-/Discipline-and-Reward/Dealing-with-a-very-angry-child.aspx

https://www.supernanny.co.uk/Advice/-/Parenting-Skills/-/Discipline-and-Reward/Calming-your-kids-how-do-you-tame-a-wild-child.aspx

http://www.unk.com/blog/treating-the-angry-client/

Derren Brown. Happy: Why More or Less Everything is Absolutely Fine Hardcover. Bantam Press. ISBN-13: 978-0593076194

No pain, no gain?

A look at ways to reduce pain.

The way that people learned how to do their job was probably a mixture of courses, working with a more experienced colleague, and trial and error. And once they became a bit of an expert, it didn't end there. They continued needing to stay up-to-date with the latest trends and developments in their area. The list of things they need to know about just continues to grow. And they're not going to gain any more knowledge without experiencing the pain of learning.

Interestingly, all pain is psychological – it doesn't exist anywhere, it's just a subjective experience in the brain. You've probably seen sports people who break a limb and carry on playing without feeling any pain. Somehow, the adrenalin coursing through their body is turning off the pain that you or I might be feeling with the same injury. Of course, the opposite is also true. If the doctor tells you that you don't have any physical injuries and there's no source for the pain, but you still feel in pain – then you are in pain!

Apart from taking drugs to reduce the pain (or other drugs to increase the pain), the amount of pain that a person feels, ie the level of the pain (the subjective units of distress), is affected by the way that the person thinks about the pain.

Depending on how a person thinks about their pain can influence how much they catastrophize about the pain. People who are prone to catastrophize, experience an exaggerated negative mental set. Of course, some people are quite stoical about pain – but it may be that they've never experienced really bad pain. Some people tend to ruminate about pain – they keep thinking about painful episodes. And some people seem to feel quite helpless about pain. They feel they can't do anything to reduce the pain or cope with it.

It appears that people with higher levels of catastrophizing about pain not only rate their own pain as more intense, but also tend to experience increased levels of disability following knee surgery. Following an injury at work, they're more likely to experience chronic long-term pain rather than short-term pain that ends when the physical injury heals. They also tend to experience higher levels of depression, they take more painkillers, and stay longer in hospital.

You might wonder how these kinds of result were obtained, and how researchers induce pain in people. They needed a way that leaves no lasting damage, so what they did was shine lasers onto the test subjects' skin, or sometimes they got them to hold their hands in a bucket of icy water.

If you are a pain catastrophizer, what can you do? It appears that a person's perception of pain can be reduced by receiving appropriate psychological therapy. An easier way to reduce pain and reduce your pain catastrophizing is distraction. And the simplest distraction is to hear a joke. Anything that takes your mind off the pain will help to reduce the pain.

Clearly there are ways that you can help the clients you're working with to deal with any pain they are struggling with more successfully.

Acute pain is designed to alerts us to danger, which it does by sending signals to the 16 parts of the brain that process pain. And that create a perception of pain. With acute pain, once the danger is over, the brain creates a 'counter signal' that dampens the pain as the tissue heals.

With chronic pain, inflammation continues, and anti-inflammatory processes are overwhelmed, resulting in the nerve cells dedicated to pain increasing by up to five times. However, we can use visualization and other non-painful stimuli (thoughts, images, sensations, memories, soothing emotions, movements, and beliefs) to activate the same parts of the thinking brain that would typically be processing the persistent pain, and force them to be used for a different purpose. And, because of neuroplasticity, those 16 parts of the brain become used to processing non-painful stimuli instead of the pain.

There is a technique that Michael Moskowitz came up with following a serious accident and suffering chronic pain. The technique is very hard because the person has try their non-painful stimuli every time they experience the pain – and the pain is pretty much constant to begin with. Eventually it begins to work. Moskowitz came up with the acronym MIRROR to describes the qualities an individual needs to succeed with this technique:

- Motivation – stay motivated to change the brain even without immediate success.
- Intention – focus on changing the brain to stop persistent pain.
- Relentlessness – counter-stimulate every pain intrusion using thoughts, images, sensations, memories, soothing emotions, movement, and beliefs.
- Reliability – count on the brain to make positive change.
- Opportunity – use pain intrusions as an opportunity to practice neuroplastic treatment approaches to stop the pain.
- Restoration – disconnect expanded pain circuits, shrink the brain pain map, and restore pain to its important role of sounding an alarm about danger.

Interestingly, Milton Erickson listed eleven methods of dealing with pain using hypnosis (see Erickson, 1980, Vol 4, p 240-245). These categories, are:

- Directly suggesting that pain will disappear.
- Indirectly suggesting that pain will disappear.
- Creating amnesia for past experience of the pain.
- Creating numbness or analgesia in the painful area of the body.
- Creating a more total anesthesia by having the person imagine they are somewhere far from the pain.
- Altering sensations of pain into sensations of itching, warmth, coolness, or other less disturbing sensations.
- Displacing the pain to a more manageable area of the body (eg moving abdominal pain to a hand).

- Dissociation, eg by having the person imagine that they are across the room observing themselves.
- Reinterpreting the pain as a feeling of heaviness, pulsation, or movement.
- Distorting time perception so that a prolonged period of pain seems to go by much faster.
- Suggesting that the pain will reduce itself very gradually; so gradually that the person cannot even monitor whether or not this is happening.

Reference

Moskowitz, M. H. & Golden, M. D. (2013) Neuroplastic Transformation: Transforming the Brain in Pain. Neuroplastic Partners, LLC.

Erickson, M. The Collected Papers of Milton H. Erickson (E. Rossi, Ed.). Somerset, N.J.: John Wiley & Sons, 1980. Innovative Hypnotherapy Vol 4.

How to stop brooding

What to do when your client is brooding and how to help them stop.

We've all had days where something bad happens – your boss shouts at you, your children shout at you, someone pushes into the queue in front of you, someone cuts you up on the motorway, etc, etc – and rather than just letting it go as an unpleasant experience (just one of those things that we can't control), we start thinking about it. We start thinking of witty one-liners that we could have used as a put-down. We think of long explanations that would have clarified our view of things. So, rather than having that bad experience once (and filling our stress bucket just a little), our brain has that experience, twice, three times, twenty times, perhaps even a hundred times before we go to bed that night. And each one of those upsetting re-runs is filling up that metaphorical stress bucket more and more. We're brooding, or ruminating, or overanalysing.

Clearly, the right pre-frontal cortex is there for a very good reason. And evaluating what has happened can be very useful. We can decide what went well and what would be better if things were different. But some people don't seem to just visit their right pre-frontal cortex for a short period of time, they seem to stay there, not just for most of one day, but seemingly every day. But apart from filling our stress bucket, brooding doesn't do any harm does it?

Unfortunately, it does. The negativity we feel about that one event can spread to negatively viewing other aspects of our lives. And that leads to depression and longer depressive episodes. Some people use alcohol as a way of stopping brooding. And so the tendency to brood is associated with a greater risk of alcoholism. Other people try to comfort eat as a way of cheering themselves up and stop brooding. So brooding is also associated with increased risk of developing an eating disorders. It's also linked to anxiety and substance abuse. Of course, the brain can't tell a real event from an imagined event (or a brooded over event), so it begins to think that your life is full of confrontation and unpleasantness. This can lead to feelings of helplessness and passivity, which makes a person less likely to take remedial action. Brooding is also bad for our health and can lead to us developing cardiovascular disease.

But what can you do? How can you stop brooding? How can you move from that right-pre-frontal cortex to the left? Just wishing it, doesn't work. So, what does?

Distraction does work. So if you feel yourself running over an event for the second or third time, try a crossword puzzle or Sudoku. You could also watch an absorbing film or TV programme (as long as it isn't on the same topic), or do a quiz. Do anything that requires concentration.

Like with so many things, exercise can help. There's a theory that the more you're thinking about a problem, the slower you walk. So if you run, then you can't really think at all.

You could reframe the event by looking upon it as a learning exercise so that you know how to respond/act another time. Or think of it as an experiment, and observe the results when this particular set of circumstances occurred.

You could take control of your life and only allow yourself to worry at a specific time of the day, say 20-30 minutes in the early evening. At other times, stop yourself brooding because it's the wrong time of day.

> Positive thoughts, positive actions, and positive interactions are thought by some to be the cornerstone of mental wellbeing.

If you see clients who have a tendency to brood on events, then any way that they find helps them to stop has got to be a good thing. And once they've stopped brooding, you can start really emptying their stress bucket (they won't be filling it up quite so quickly), and getting them to identify sparkling moments (good things that happen in their life) and things they are grateful for.

References:

http://www.huffingtonpost.com/guy-winch-phd/negative-attitude-and-health-_b_3749557.html

https://psychcentral.com/blog/archives/2014/02/16/8-tips-to-help-stop-ruminating/

http://confidencecues.com/3-secrets-on-how-to-stop-brooding-over-the-past/

How to perform under pressure

A look at what helps people to perform well under pressure.

There must be times when you feel like a juggler with too many balls in the air, you're not sure that you're going to be able to catch them all and throw them again successfully. Certainly you're going to see clients who have come to you because of the stress in their life. Apart from helping them relax and empty their stress buckets, as well as getting them back into their intellectual brain, what else can you do for them?

Before we go any further, let me ask you a question: are you an optimistic person? Various studies have shown that optimists do better at school, in sport, and at work. In a study of insurance salespeople, the most optimistic people sold 88 percent more than the least optimistic people. Optimists have better health and live longer. Optimists look on bad events as temporary rather than permanent. They make specific rather than universal explanations for bad events. And they attribute bad events to external causes rather than internal ones, eg the computer crashed because of bad code rather than because I am a bad user.

So, what can our clients do to perform well under pressure and get this problem solved? The first thing is to learn to interpret their body signals differently. The ABC method was devised by Albert Ellis in the 1950s. A stands for activating event, ie the triggers that make you feel scared. B stands for beliefs, ie your assumptions about this situation. And C stands for consequences, ie the feelings and behaviours that result from A and B. If the result isn't what you want, you can move through the alphabet to D, which stands for dispute. This looks for alternatives to your current beliefs. And E, which stands for energizing alternatives. This is where you look for more empowering beliefs to adopt.

The second way of performing better under pressure is to go through a pre-performance routine. This is where you force yourself to focus on carrying out a set pattern of activities. While you are doing this routine, you are stopping yourself from focusing on any potential negative thinking. This works particularly well before giving that all-important presentation or going into negotiations (eg for a pay rise at work).

The third technique to use is positive self-talk. If your client says to themself that this particular problem is going to lead to an extinction-level event, their company will be out of business within a couple of weeks, and, whatever happens, they're going to lose their job, this is going to send their anxiety levels sky high. Whereas, if they say to themself that they have always been able to sort out these kinds of situation in the past and there's no reason why they won't be able to get on top of this one today, then their anxiety levels won't increase and may well go down.

Technique four is to visualize themself succeeding. This works equally well on the golf course, or when giving a presentation, or in many other situation. They know the sorts of thing they need to do, so they just need to picture themself doing those things – using their computer to gain more information, phoning people to turn on or off different components, whatever it is that they usually do. Maybe, they're walking to a different part of the building or using their tablet to get to the nub of the problem. As part of the visualization, they need to see what they're likely to see, hear what they might hear, and feel whatever they might feel.

And remember, it's always good to breathe, so they can try 7-11 breathing, where they breathe in to the count of seven and breath put to the count of 11. If that proves hard, then they simply breathe out for longer than they breathe in. This will have a calming effect on them.

Although in this situation it's too late, but having a good night's sleep makes them better able to handle stressful situations. Lack of sleep can leave people cognitively impaired. If you know that tomorrow you have to give an important presentation (or you have an important golf match) then a good night's sleep is important. If you can't get to sleep, then try counting backwards from 300. If your mind drifts off, start again at the last number you can remember counting down to.

Nothing makes handling difficult situations easier than practice. The more times you have dealt with problems, the easier it is to know what to do. That's why fire crews and the army are always rehearsing what to do in life-threatening situations. Your client's situation may be bad, but it's probably not life threating. But the more they rehearse dealing with situations (and that can just be in their head or on a piece of paper) the better able they will be to perform well under pressure.

References:

Alison Price, David Price. Psychology of Success: Your A-Z Map to Achieving Your Goals and Enjoying the Journey. Publisher: Icon Books Ltd. ISBN-10: 1785780212

Motivating your clients

Getting your clients motivated to make positive changes in their life can be very similar to them going shopping.

You may wonder what motivating your clients and shopping have in common. The answer is dopamine. Dopamine is a neurotransmitter that helps carry a message from one neuron to another across the gap (or synapse) between them. It also acts as a neuromodulater. This is where one neuron secretes a neurotransmitter that diffuses through large areas of the brain and affects lots of other neurons.

The hypnotherapy sandwich

You might have heard the phrase hypnotherapy sandwich, but what is it? Well, of course, it's a metaphor – isn't that what we use all the time? It's a way of reminding us what to do at the start of the session and at the end.

The point of the sandwich is that it:

- Focuses on the client's desired outcome.

- Results in a logical conclusion to the session that empowers the client.

- Puts the onus of continued change on the client (where it belongs).

The first question to ask at the start of the session is: "What do you want from hypnotherapy today?" Focus on what they do want, not on what they don't want – that list could be huge and is problem focused not goal focused.

At the end of the session, you can put the responsibility for continued progress back on the client, by asking: "What's next?". This gently but positively nudges the client to focus on their next steps.

When you're shopping, or browsing your way through a Web site, your brain is constantly giving little dopamine surges as you view each product. Some surges are bigger than others, and the difference reflects the expected pleasure that each item will give you. The more you want something, the more dopamine you produce. Or, the more dopamine you produce, the more you want something. And when you want something a lot, you continue to want it after you have left the shop or the Web site. So, dopamine keeps on encouraging you to do something – in this case, buy something. In the client's case, it could be going to the gym.

Dopamine, also kicks in when you get pleasure from doing something. So a success or a win will result in dopamine being produced and you continuing to engage in whatever activity you were doing that resulted in the success or win. It's motivating. This could be when your client enjoys running or using a particular piece of gym equipment.

And going back to the shopping analogy, when you open the parcel with your expensive purchase in it, there's another dopamine surge, making you feel so pleased that you bought that particular item. This is where hedonic adaptation comes in. Hedonic adaptation is a term used to describe the observed tendency of people to quickly return to a relatively stable level of happiness despite major positive or negative events or life changes. So after you unpacked your new purchase, you gradually feel less-and-less thrilled about it and take it for granted. You sort of acclimatize to it. This means that clients will begin to feel less motivated to go to the gym the more often they go because they will be getting less of a dopamine hit. And that's where we need to keep encouraging them.

The dopamine messages also stop if the item you would like is totally out of reach. There's no point spending hours looking at brochures for yachts if you can barely afford to pay the rent on the property you currently live in. Or there's no point having a goal of losing ten stone when the client can barely shed a couple of pounds because of their life style.

So dopamine has a big effect on clients who want to make positive changes to their life. When they feel strongly that they want to achieve their goal, that's the dopamine having an effect. The more they feel that they want to achieve their goal, the more powerful their dopamine hit will be and the more it will focus their attention on achieving their goal. As well as motivating a person, dopamine can make someone addicted to trying to get the reward from achieving their goal. That really is a strong driver for getting your client to make those changes.

One way to utilize this is give your client some quick wins – some goal-related tasks that aren't too difficult or take too long. And, of course, celebrate their success when they report back to you. In fact, you might want to break down the task into separate smaller tasks. That way the client can get multiple successful hits of dopamine as they work their way through to completing the bigger task. If you find that your client is losing motivation along the way, it may be that these subtasks are too big and the rewards are not arriving quickly enough to keep the dopamine coming. The trick then is to further subdivide the tasks or add new smaller ones that can be accomplished while working on the larger task.

As we've been talking about the brain, it's worth noting that when you're learning to do something, like ride a bicycle or drive a car, you are using your pre-frontal cortex. This uses lots of energy and, while being very logical, is also slower to access than other parts of the brain. Once we've done a task a few times, it becomes a habit, we can do it without giving it any conscious thought. You may have driven home from work without thinking about driving at all. An area near the middle of the brain, called the basal ganglia, runs things on autopilot, freeing up the rest of the brain to concentrate on other things. Your client will eventually be able to perform whatever new task (new habits) they are currently trying to learn almost without thinking about it utilizing their ingrained habits.

Shopping is something we all have experience of doing, and some people enjoy it more than others. It may well be that people who enjoy it the most buy the most products. Understanding the shopping analogy can help to encourage clients to start new positive habits and motivate them to become as successful as the experts at whatever it is that they want to achieve (eg a particular sport). Knowing about dopamine makes it clearer how to bring on your clients to expert levels of success.

The other thing to bear in mind is that, boadly, there are two types of motivation – intrinsic and extrinsic. Intrinsic motivation to do something comes from your own interests and enjoyment. Extrinsic motivation is driven by external incentives and disincentives, such as money, high marks, competition, coercion, and fear of punishment. People who are self-motivated (ie their motivation is intrinsic) enjoy a number of benefits. They have:

- Greater confidence
- Improved vitality
- More interest
- Enhanced performance
- Increased persistence
- Greater creativity
- Higher self-esteem
- Increased general wellbeing.

Your client will set their own goals. There are, basically, two types of goal – approach goals and avoidance goals. Approach goals have positive outcomes and they are goals that people are working towards, eg be able to get on a bus on my own. Avoidance goals have negative outcomes and are usually things that a client is moving away from. An example might be to cut down on the number of cigarettes smoked in a day.

Psychologists believe that approach goals are better ones to set because they can contribute to a person's wellbeing. (You may remember seeing old newsreels of anti-war marches. Nowadays, they are always peace rallies. The emphasis has changed from an avoidance goal to an approach goal.) Making progress towards achieving a valued goal makes us feel good. Avoidance goals can be stressful because we are constantly monitoring negative possibilities. Avoidance goals lead to the absence of something negative. Approach goals lead to the presence of something positive.

Edward L Deci and Richard Ryan came up with a Self-Determination Theory (SDT) of motivation. They thought that three basic human needs had to be met in order for a goal to be met. The three basic needs are autonomy, competence, and relatedness. Autonomy is where you are in control of what you do and when you do it. Autonomy is good for self-motivation and wellbeing. Being forced to act in a particular way has a negative impact on self-motivation and wellbeing. Competence is the human need to feel confident, effective, and masterful. Relatedness is the need to have relationships with other people, ie to be connected to other people.

Sonja Lyubomirsky says that goals that are intrinsic, congruent with your motives and needs, and not in conflict with each other are likely to enhance a person's happiness and life satisfaction.

References:

Alison Price, David Price. Psychology of Success: Your A-Z Map to Achieving Your Goals and Enjoying the Journey. Publisher: Icon Books Ltd. ISBN-10: 1785780212

Alison Price, David Price. Psychology of Success: Your A-Z Map to Achieving Your Goals and Enjoying the Journey. Publisher: Icon Books Ltd. ISBN-10: 1785780212

He who laughs, lasts

A look at laughter and happiness.

We're all familiar with the ups and downs of daily life, but does it make you happy? Perhaps the best branch of psychology to use to see whether hypnotherapists are happy at work is positive psychology. Positive psychology uses scientific understanding and effective intervention to aid in the achievement of a satisfactory life. Its focus is on personal growth. According to positive psychology, happiness is improved and affected in a large number of different ways, for example: social ties with a partner, family, friends, and wider networks through work, clubs, or social organizations. As we suspected, happiness increases with increasing financial income, but it reaches a plateau and no additional pay rises make you any happier. It's also worth noting that physical exercise correlates with improved mental wellbeing.

As well as helping people (our clients) change their negative style of thinking about other people, their future, and themselves, positive psychology also helps families and schools to allow children to grow; and it can be used to create workplaces that aim for satisfaction and high productivity. I guess that we've all met other therapists (never us, of course), who seem to be negative about everything. They're not working in an environment designed using positive psychology principles. Clinic managers like positive psychology – what manager doesn't want productivity and success rates to be as high as possible?

Positive psychology focuses on: positive emotions (being content with your past, being happy in the present, and having hope for the future), positive individual traits (your strengths and virtues), and positive institutions (strengths to improve a community of people). One problem people often have with work is remembering the good parts. We've all had clients like this – most of the day was good except for an hour in the afternoon. And that's the bit they remember and tell people! One way to get an accurate record of how clients feel during a typical day is to have people (scientists) use beepers to remind them to write down the details of how they currently feel – hopefully not irritated because a beeper has just gone off! Basically, this illustrates the difference between the 'experiencing self' and the 'remembering self'. Daniel Kahneman identified a cognitive bias that he called the peak-end effect. What that means is that people remember the dramatic parts of a day and the end. So try not to let your clients leave a therapy session without giving them a few minutes of a pleasant experience – particularly on, what for them has been, a bad day, because that will colour how they remember the whole day. Generally, spending time in trance is one of the high spots of any client's day.

Martin Seligman came up with the acronym PERMA (Positive emotions, Engagement, Relationships, Meaning and purpose, and Accomplishments) for wellbeing. Positive emotions include happiness, joy, excitement, satisfaction, pride, and awe. Engagement refers to involvement in activities that draw and build on a person's interests (what Mihaly Csikszentmihalyi called 'flow'). It involves passion for and concentration on the task at hand. Relationships are about receiving, sharing, and spreading positivity to others. Meaning (or purpose) drives people to continue striving for a desirable goal. Accomplishments are the pursuit of success and mastery, which, strangely, may not result in positive emotions, meaning, or relationships.

You can see that PERMA can apply to clients during their working day. They may well have 'positive emotions' about their job (or certain aspects of it). They could spend large parts of the day (providing the phone doesn't ring too often) in the 'flow' – totally 'engaged' in their work. When they're solving a problem and they're completely absorbed in the task; when they're using all their years of knowledge and experience to identify the solution – that's being engaged. And when they're chatting to co-workers positively about work-related matters, that's ticking the 'relationships' box. 'Meaning' is what drives people to achieve their goal. And finally 'accomplishments' makes people study to pass exams to learn more about whatever job they do.

If our clients did measure, every ten minutes, how happy they were, they'd probably find that the experience of each day was actually quite different from how they usually remember it.

Interestingly, when people enjoy their work and laugh out loud, they're going to work together better, they're going to come up with better ideas, and it's going to be healthier. Let's have a look at what's going on when you and your clients are laughing.

In terms of emotional intelligence, laughter helps people view situations in a more realistic and less threatening light. This change of perspective can make your clients, in their work environment, more empathic and better able to understand the points of view of their work colleagues. It appears that laughter can strengthen relationships and enhance the way people work as a team. It also seems to promote group bonding and help to defuse conflict situations. And (although this one isn't so useful in a working environment) it makes us more attractive to others! Laughing also helps to increase a person's resilience to stress (and there's plenty of that in the workplace) and helps them to find meaning in their life. It can also help people create a positive outlook that can be applied to all aspect of their life (including work) – definitely lifting what might be described as a low mood.

It also seems that laughter can help people think 'outside of the box' and be more innovative and creative – coming up with new ideas that could help their organization be more competitive. Not only do people become more emotionally aware, laughing also improves their memories. The hormone cortisone, which is produced in moments of stress, can damage the neurons in a person's hippocampus and can even shrink the size of their brain. So, laughter reduces the amount of cortisone and helps with memory. In fact, one study found that people who laughed were able to learn and recall information in almost half the time of those people who didn't laugh!

Laughter is good in so many ways. For example, laughing relaxes a person's whole body. The act of laughing increases abdominal pressure and movements of the diaphragm. These movements massage the vagus nerve, causing it to send a signal telling the body to relax (using parasympathetic nerves). The body movements that go with laughing also act like a pump for a person's lymphatic circulation. This assists the lymphatic vessels in carrying fluid through that person's body and helps their lymph nodes to clean and filter this fluid, removing waste products, dead cells, and even unwanted microorganisms. An increased lymphatic flow improves their immune system. As well as decreasing the levels of stress hormones (cortisone), laughter increases the numbers of immune cells (lymphocytes) and infection-fighting cells (phagocytes), and so improves a person's resistance to disease and ability to fight infection. It also causes the body to release endorphins. These act as pain killers and promote an overall feeling of wellbeing.

There's evidence suggesting that a hearty laugh relieves physical tension and stress, leaving a person's muscles relaxed for up to 45 minutes afterwards. Also, when a person laughs, they stretch muscles throughout their face and body. This results in their pulse and blood pressure going up, and they breathe faster, sending more oxygen to their tissues. This can increase a person's energy levels and make them more productive at work. Laughter also, apparently, causes the release of nitric acid, which helps dilate blood vessels, which, in turn, protects your heart.

Laughter has been shown to help hospital patients with a range of illnesses, making them better able to cope with their illness and their treatment. It's also very difficult to feel angry, anxious, or sad if we are laughing. Laughter helps us keep a positive, optimistic outlook when we're experiencing difficult situations, disappointments, and losses.

As well as helping us create endorphins, laughter affects our opioid system and both of those are associated with stress-induced emotional eating. 10-15 minute of laughing burns 50 calories (according to a 2015 study conducted by Maciej Buchowski, a researcher from Vanderbilt University). So laughter helps with emotional eating problems. Laughter can also reduce blood sugar levels. There was a study of 19 people, who ate a meal and then sat through a tedious lecture – after which they had their blood sugar levels measured. The next day, they ate the same meal and watched a comedy – and had lower blood sugar levels than the previous day.

So, if laughter came in tablet form, we'd be queueing up to buy it and give it to our clients as a supplement. Laughter makes people feel better, work better, and relate to each other better. Therapists should (and usually are) encouraging all their clients to laugh out loud.

References:

Bridget Grenville-Cleave. Positive Psychology (A Toolkit for Happiness, purpose, and wellbeing). Icon Books Ltd. 978-1848319561.

Time for a change

Here is an overview of an effective NLP technique to help people control their eating.

The essence of all therapy is to use the best tool for the job. Each type of therapy focuses on one area of work – such as acupuncture, EFT, or Reiki – and each has a degree of success. The important thing, for our clients, is to use the best tool possible to bring about positive and lasting change. But, in carpentry terms, "to a man with a hammer, everything looks like a nail". And that's why knowing what works is important, and how well it works when compared to other 'tools' in your metaphorical 'toolbox'.

NLP has provided us with tools and techniques such as reframes (the fast phobia cure), swish, anchoring, and circle of excellence. There's another really effective technique from NLP that can help people who, for example, over-indulge in chocolate to stop eating it completely. From there, that person can go on to get control of their eating and their weight. So how does it work?

The idea is to replace the submodalities of something you really like and want to stop eating, eg chocolate, with the submodalities of something you never eat, eg spinach. But what are submodalities? We have five basic senses: visual, auditory, kinaesthetic, olfactory, and gustatory. In NLP, these are referred to as representational systems or modalities. And each of these can be divided into smaller components.

Visual submodalities include:

- Style: picture, painting, poster, drawing, 'real life'
- Still or moving
- Panoramic versus framed picture
- Shape: concave, convex, square, oval, etc
- Level of brightness
- 2D versus 3D
- Black and white versus colour
- Focus (on certain objects) versus blurred
- Clear versus fuzzy
- Movie versus still frame
- Size of the picture (tiny, small, life size, big, huge)
- Size of the main object
- Associated (seeing through your own eyes) versus dissociated (seeing yourself in the picture)
- Number of pictures viewed simultaneously
- Distance of the picture from you (near or far)
- Sharpness of colours
- Contrast

- Angle
- Movement (stopped, slow, regular, fast, super speed)
- Location of picture (up, down, left, right)
- Bright or dim
- Clear or grainy
- Solid or transparent.

Auditory submodalities include:

- Mono versus stereo
- Loud or quiet
- Inflections (words marked out)
- Pauses
- Duration
- Rhythm (regular, irregular)
- Volume
- Variations: looping, fading in and out, moving location
- Tonality
- Qualities of sound (raucous, soft, windy)
- Static versus moving
- Location/direction
- Tempo
- Soft versus rasping
- Frequency (high versus low pitch)
- Source of sound
- Cadence
- Timbre (characteristic sound, such as a voice like Bugs Bunny)
- Movement of the source
- Tempo
- Voice: whose voice, one or many
- Background sound versus only sound
- fast or slow
- near or far
- single sound or layers of sound
- a tone, a voice, musical etc
- speed or duration

- pauses in the sound.

Kinaesthetic modalities include:

- Temperature (hot versus cold)
- Texture (rough versus smooth)
- Vibration
- Pressure
- Weight (heavy versus light)
- Location
- Rhythm
- Steady or intermittent
- Facial expression
- Body position
- Eye positions
- Gestures
- Mass – how big is it?
- Intensity
- Density
- Movement (motion, spinning in which direction)
- Balance
- Strong or weak
- Constant or intermittent
- Shape of sensation
- Where the feeling is in your body
- Feelings shape or a texture.

So, what do you do? You map the submodalities of the spinach onto chocolate, and the client will then hate chocolate.

Firstly, elicit the visual, auditory, and kinaesthetic submodalities of chocolate. So, you ask:

- Do you have a picture?
- Is it black-and-white or colour?
- Is it bright or dim?
- Near or far?

Use other submodalities from the list, and write down the answers. Then do the same for sound and feelings. Break state after asking the questions (by asking another question like what's your phone number). Breaking state is an NLP term for changing someone's emotional state. It's used when you want to break someone's concentration by switching their attention from one thing to another. The next stage is to elicit the submodalities of spinach. And finally, compare the two sets of submodalities to find the difference that makes a difference (another NLP phrase!).

Some of the differences between the two lists will have a more powerful influence than the rest, and these are called the drivers. Usually, changing a driver will cause all the other submodalities to change with it automatically. As a general rule, the three things that stand out as drivers are:

- The location of the picture
- The size of the picture
- The difference between associated and dissociated.

The final stage is to map the submodalities of spinach onto chocolate.

So, let's assume that the client's picture for chocolate is a large and colourful in the centre of their field of vision and is dissociated (and by that I mean that they can see themselves eating chocolate and enjoying it). Whereas their picture for spinach is smaller, in the bottom-right corner of their field of vision, is black and white and associated (by that I mean that they can see a forkful of spinach coming toward their mouth).

We then ask them to imagine a dissociated, black and white picture of a chocolate coming towards their mouth and have them move that picture into the bottom-right corner of their field of vision and make it the same size as the spinach picture. As a result of this, the client will find that they dislike chocolate with the same intensity that they dislike spinach.

You can do things in reverse, too. So if a client wants to like something, you can change the submodalities for that thing to something that they like a lot.

This is another useful tool to add to your toolbox.

Socratic questioning

This is a look at how asking your client questions can help to change their thinking and their behaviour.

Before we start looking at Socratic questioning, it probably makes sense to see who Socrates was and what kind of questioning he invented.

Socrates, who lived from 470/469 to 399 BCE, is credited with being one of the founders of Western philosophy. What we know about him comes mainly from his student, Plato.

Socrates came up with the Socratic method, which he used largely to examine concepts such as 'Good' and 'Justice'. To solve a problem, it would be broken down into a series of questions. These questions are posed to help a person determine their underlying beliefs and the extent of their knowledge. Better hypotheses are found by steadily identifying and eliminating those that lead to contradictions.

That all sounds very useful if you're a philosopher, but how does it help with clients? The answer is that it can be used as a cognitive restructuring technique, where it helps to uncover the assumptions and evidence that underpin a client's thoughts about whatever it was that brought them to see you.

By using Socratic questioning, you can challenge recurring or isolated instances of a client's illogical thinking while maintaining an open position that respects the internal logic to even the most seemingly illogical thoughts. Socrates thought that people already had some knowledge, so rather than teaching them, he would draw out what they already knew, and they would act on this pre-existing information. And that's the bit that can make it so useful, particularly with clients with depression, and with clients who are catastrophizing and using all-or-nothing thinking. You can't tell them they're wrong, but they can realize from their own knowledge that there is an alternative way of seeing things. And that makes it more acceptable to them. It's a way of changing their mindset, a way of getting them to see the big picture – it's, in effect, a reframe. It's a great way to get over a client's emotional biases, which are resistant to any kind of logical argument.

Rather than listing example questions, I'll divide them up into 'types' of question.

The first type are the questions that try to reveal more about what the client is thinking about – trying to clarify things. These are 'tell me more' questions that examine the concepts behind their argument. You might ask questions such as:

- Why are you saying that?
- What exactly does this mean?
- How does this relate to what we have been talking about?
- What is the nature of ...?
- What do we already know about this?
- Can you give me an example?
- Are you saying ... or ... ?

- Can you rephrase that, please?
- What evidence supports this idea?
- What evidence is there against its being true?
- Where is the evidence for your belief?
- Are there any exceptions to that?
- Is that belief completely rational and logical?
- Is there a valid reason that you believe that statement?
- Is this logical? Are there any errors you might be making in your thinking?

The second type of question attempts to probe their assumptions making them think about the presuppositions and unquestioned beliefs on which their thoughts are founded:

- What else could we assume?
- You seem to be assuming ... ?
- How did you choose those assumptions?
- Please explain why/how ... ?
- How can you verify or disprove that assumption?
- What would happen if ... ?
- Do you agree or disagree with ... ?

The third type of question digs into the reasoning behind their argument and the evidence, rather than assuming it is a given. This picks up on the fact that people often use un-thought-through or weakly-understood ideas for their arguments:

- Why is that happening?
- How do you know this?
- Show me ... ?
- Can you give me an example of that?
- What do you think causes ... ?
- What is the nature of this?
- Are these reasons good enough?
- Would it stand up in court?
- How might it be refuted?
- How can I be sure of what you are saying?
- Why is ... happening?
- Why? (keep asking it – you'll never get past a few times)
- What evidence is there to support what you are saying?
- On what authority are you basing your argument?

The fourth type of question assumes that most arguments are given from a particular position. So, it attacks the position and attempts to show that there are other, equally valid, viewpoints or perspective.

- Another way of looking at this is ..., does this seem reasonable?
- What alternative ways of looking at this are there?
- Why is ... necessary?
- Who benefits from this?
- What is the difference between ... and ...?
- Why is it better than ...?
- What are the strengths and weaknesses of ...?
- How are ... and ... similar?
- What would ... say about it?
- What if you compared ... and ... ?
- How could you look another way at this?
- What might be another explanation or viewpoint of the situation?
- Why else did it happen?
- Can you think of any circumstances where that idea simply doesn't hold true?
- Are you making an over-generalization there? Why?
- Does your idea apply in all contexts, and if not, why not?
- Would your friends and colleagues agree with that statement?
- Does everybody share your attitude? If not, why not?
- If you were to challenge or criticize that statement, what would you say about it?
- What alternative views are there? What is the evidence for them?
- Is it possible that....
- Are there other reasons why x did y?

The fifth type of question assumes that the argument the client gives may have logical implications that can be forecast. Do these make sense? Are they desirable?

- Then what would happen?
- What are the consequences of that assumption?
- How could ... be used to ... ?
- What are the implications of ... ?
- How does ... affect ... ?
- How does ... fit with what we learned before?
- Why is ... important?
- What is the best ... ? Why?

- What are the worst, best, bearable, and most realistic outcomes?
- What is the effect of thinking or believing this?
- What could be the effect of thinking differently and no longer holding onto this belief?'
- Where is this thought/attitude/belief getting you?
- Is this thought/attitude/belief helping you to achieve your outcome?
- What are the consequences of telling yourself that?
- Is there a more constructive way of dealing with the situation?

The sixth type are questions about questions:

- What was the point of asking that question?
- Why do you think I asked this question?
- Am I making sense? Why not?
- What else might I ask?
- What does that mean?

Other Socratic questions can be used help to distance (dissociate) the client from their feelings:

- Imagine a specific friend/family member in the same situation, what would you tell them?
- If a specific friend/family member viewed the situation this way, what would you tell them?

To summarize, Socratic questioning can be used in your consulting room as a way to change how you client views the world. This then changes their emotions and their behaviour – which is what they want and why they came to see you in the first place. You're not trying to get them to accept new ways of thinking based on what you say, rather you're changing their thinking by using what they already know. You're bringing it to the foreground of their mind. And you're doing it all by asking them questions.

References:

http://www.sciencedirect.com/science/article/pii/S0005796715000790

http://www.unk.com/blog/socratic-questioning-in-depression-therapy/

http://www.adam-eason.com/using-socratic-questioning-techniques-in-hypnotherapy/

https://en.wikipedia.org/wiki/Socratic_questioning

http://changingminds.org/techniques/questioning/socratic_questions.htm

Successful clients

Let's take a look at ways of getting your clients to do their best.

What separates those clients who make amazing progress from those who seem to be almost happy with their problem and don't seem to want to change? The answer could well be how motivated they are to improve their life.

To start with, let's take a look at what makes a top performer different from an average performer. A study was carried out on musicians at a music school by Anders Ericsson. With help from the tutors, he divided the students into three groups. The first group contained students that it was thought would never make it as a musician and were doomed to teach music. The second group were musicians who it was thought were good enough to become professional musicians. The third group contained students who were thought to be so good that they could become world-class musicians. The psychologists then looked for what factors affected which group students ended up in. And the one factor that seemed to determine their group was the amount of time they spent practising. Group one did 4000 hours of practice. Group 2 did 8000 hours of practice, and the top group did 10,000 hours of practice.

So, if you want your clients to make outstanding progress, they need to do their homework – they need to listen to the audio track/CD, they need to look out for sparkling moments in their week, and they need to think about what was said. And the more practice they do, the more progress towards their goals they make. The study of musicians couldn't find any sign of innate talent. The only correlating factor was time spent practising. And it's not just some practice that's needed, it's lots and lots. It's that dedication to practising that makes all the difference. And that's why, for their own good, you should insist that clients do their homework.

The same is true for sports people. Michael Phelps is an American swimmer, who is the most decorated Olympian of all time, with a total of 28 medals. To get to that level of ability, Phelps trained almost every day for 12 years. Phelps set himself the goal of doing what no-one else had ever done before. This is called the effort heuristic by psychologists. It's the amount of effort that someone will put into trying to achieve something, as determined by the value they place on that goal. The effort that someone puts in to their homework (and listening to an audio track as you drift off to sleep isn't hard!) is equal to the reward that they expect to get as a result of that effort, eg feeling less depressed or not being scared of spiders, or sleeping through the night. So, how successful someone is (whether that's in music, sport, or achieving their personal goal) depends on how much they want to achieve their goal.

And that means for you, that you need to make it crystal clear to your clients exactly what rewards they can expect from hypnotherapy, so that they are motivated and enthused to achieve that goal.

Now, let's imagine a work situation. Let's suppose that two new people start working for an organization and they have two different line managers, who have slightly different ways of working. One of the managers is very keen on SMART (Specific, Measurable, Achievable, Relevant, and Time bound) targets, and the other prefers a more relaxed approach where staff can work at their own pace and they're encouraged to do their best. Which member of staff will get on better?

Interestingly, the answer comes from logging! Back in the 1970s, two groups of loggers were treated identically, except for one thing. One group was told to do their best and cut down as many trees as they could. The other group were taught to calculate the maximum number of trees that they could cut down in a day, and they were given a counter on their belt so that they could keep track of their progress towards the goal. The study ran for 12 weeks, after which time the goal-setting group were found to have cut down significantly more trees than the 'try your best' group. It was concluded from this (and other studies) that getting feedback on your progress towards a goal, through regular measurement, is a fantastic strategy to increase achievement.

Now, if a person were cutting down trees, it's fairly obvious what sort of indicators should be used to measure their progress. It's perhaps harder to identify indicators to use with clients, but the easiest would be subjective units – asking clients on a scale of 1 to 10 where they are in terms of happiness or depression, or willingness to pick up a spider. You can see why schools have regular maths and English tests because it provides pupils with useful feedback on how well they are progressing. You don't want to set any old indicators of progress, they've got to be the right ones, the most appropriate ones for whatever the goal is.

There's an old saying in management circles that if you can't measure something, you can't manage it. The same is true with progress towards a goal. A 2008 study looked at weight loss. 1685 overweight and obese people took part. They found that the average amount of weight lost was 13lbs (5.9kg). They also found that the average amount of weight lost by people who didn't keep a food diary was 9lbs; whereas people who did keep a food diary for at least six days a week lost an average of 18lbs.

It's definitely useful to keep a record of how clients are feeling in each session. That way, they can see how much progress they're making and they can change their behaviour accordingly so that they make more progress towards their goal.

References:

Alison Price, David Price. Psychology of Success: Your A-Z Map to Achieving Your Goals and Enjoying the Journey. Publisher: Icon Books Ltd. ISBN-10: 1785780212

Decisions decisions

This is a look at how people make decisions, and how it helped a recent client.

What can you do when you see a client who can't make decisions? His wife had recommended that Robin (not his real name) came to see me. And, according to him, most of the time he did what other people said or else he just couldn't make a decision. If he went out to dinner, he would order whatever the other person was having. He just couldn't make his mind up. He couldn't decide on what colour he wanted his new car to be, and he certainly couldn't choose the make and model!

According to Wikipedia: "decision-making is regarded as the cognitive process resulting in the selection of a belief or a course of action among several alternative possibilities". Which decision a person makes will depend on their values and preferences when they make the decision.

> Not making a decision is, itself, a decision.

So how do people usually go about making a decision? Well, there's:

- Acquiescing – agreeing with an expert or person in authority.
- Anti-authoritarianism – choosing the option that's opposite to what an authority figure might suggest.
- Balance sheet/Prioritizing – where you write down the pros and cons of each choice and choose whatever gives you the most benefit.
- Feeling (or emotional) – choosing whatever feels the right thing to do at the time, irrespective of what the consequences might be. This can be justified by rationalizations afterwards.
- Heuristic – using a rule-of-thumb for making that kind of decision based on previous experience.
- Luck – choosing an option dependent on the toss of a coin or the cut of a deck of cards.
- Satisficing – choosing the first alternative that's satisfactory rather than waiting to see whether there's a better one later.

There are some people who are quick at making decisions, but, unfortunately, they always seem to make bad ones. And they never seem to learn from their previous experiences.

In some situation, people use decision trees. These are like flowcharts, where the consequence of one event leads either to another choice or to a decision. These are often used in high-stress situations (like fire fighting or operating theatres) to make decisions that don't ignore important information that could otherwise result in a tragedy.

According to research from Washington State University, a lack of sleep slows down a person's decision making in crisis situations. So, the more sleep they get, the better they are at making decisions. And that was one of the first areas that I discussed with Robin.

Another issue that stops people coming to a decision is they feel they have insufficient information. And they continue to seek out extra information so that they can make the right choice. The trouble is, how much information is enough? And how do you know you have enough? Life isn't made up of black or white decisions. And so they never feel that they are in a position to choose. This is often referred to as analysis paralysis – over-analysing (or over-thinking) a situation so that a decision or action is never taken.

And there are some people, who feel they need their choices to be ratified by other people. They need a teacher or parent-figure to tell them that the decision they made is the right one. Or else they don't make decisions.

To help adolescents make decisions, Leon Mann and colleagues, in the 1980s, came up with a decision-making process called GOFER. The five steps in the process are:

- Goals – survey values and objectives.
- Options – consider a wide range of alternative actions.
- Facts – search for information.
- Effects – weigh the positive and negative consequences of the options.
- Review – plan how to implement the options.

Pam Brown, in 2007, divided the decision-making process into seven steps:

- Outline your goal and outcome.
- Gather data.
- Develop alternatives (eg by brainstorming).
- List pros and cons of each alternative.
- Make the decision.
- Immediately take action to implement it.
- Learn from and reflect on the decision.

Psychologists know that visceral states such as hunger and desire can affect the decisions people make. Mirjam Tuk and colleagues had people complete several different decision-making tasks after drinking a great deal of water (meant to increase the urge to urinate). The researchers found that the stronger people's urge to go to the toilet, the better they were at controlling their impulses. They concluded that the next time you are trying to make a difficult decision, in which the best option may not be the most immediately rewarding or appealing, you'll make the best decision when you are desperate for the loo!

Similarly, Kristina Guo in 2008 came up with the six parts DECIDE model:

- Define the problem.
- Establish or Enumerate all the criteria (constraints).
- Consider or Collect all the alternatives.
- Identify the best alternative.
- Develop and implement a plan of action.
- Evaluate and monitor the solution and examine feedback when necessary.

There's also BRAIN, which stands for:

- Benefits – what are the benefits to going ahead with this decision?

- Risks – what risks are associated with making this decision?

- Alternatives – what alternatives are available?

- Intuition – how do you feel about each choice? What does your gut tell you?

- Nothing – what are the consequences of doing nothing at this stage?

I quite like that one and it's actually used with hypnobirthing mums.

Or there's the DARE decision making model, where the letters stand for:

- Define the problem.

- Assess your choices.

- Respond (ie make a decision).

- Evaluate (decide whether or not you made the right decision).

Then there's the GREAT decision making model, which stands for:

- Give thought to the problem.

- Review your choices.

- Evaluate the consequences.

- Assess and choose the best one.

- Think it over afterward.

The OODA Loop and decision making

The OODA Loop refers to the decision cycle of observe, orient, decide, and act, developed by John Boyd. Boyd applied the concept to the combat operations process. It's now used in litigation, business, law enforcement, and military strategy. The approach favours agility over raw power in dealing with human opponents in any endeavour. It explains the nature of surprise and shaping operations in a way that unifies Gestalt psychology, cognitive science, and game theory in a comprehensive theory of strategy. Utility theory (the basis of game theory) describes how decisions are made based on the perceived value of taking an action. The OODA Loop shows that prior to making a decision (the Decide phase), a person will have to get information (Observe) and determine what it means to them and what they can do about it (Orient). In this way, the utility sought at the Decide phase can be altered by affecting the information the opponent receives and the cognitive model they apply when orienting upon it, when they Act.

Decisive: How to Make Better Choices in Life and Work by Chip and Dan Heath suggested the WRAP acronym for making better decisions. The letters stand for:

- Widen your options, by investigating the full range of choices available.

- Reality-test your assumptions by looking for contrary evidence.

- Attain distance before deciding. Sleep on it, and ensure your long-term goals and values are taken into account.

- Prepare to be wrong. Making ambiguous choices based on incomplete information means we will get it wrong some times. Learn from previous mistakes.

The next acronym I found is DODAR, which stands for:

- Diagnose what the problem is and the possible causes.

- Options – what choices are available?
- Decide which option to go with.
- Act or Assign – carry out selected option and/or assign tasks to appropriate people.
- Review – ensure that everything is going to plan, and the expected outcome is likely. If not, start the process again.

Similarly, you can use FORDEC as an acronym to help decision making, where:

- Facts – what is the problem?
- Options – what choices are available?
- Risks/Benefits – what is the downside of each option, and what is the upside?
- Decide which option to go with.
- Execute – carry out the selected option and/or assign tasks to appropriate people.
- Check what worked and what else needs to be done.

Another attempt at identifying all of the various activities that must occur for a decision to be made well and distilling it into an acronym is RAPID, which was originally developed by Bain & Company. It stands for:

- Recommend – someone must recommend that a decision is made.
- Agree – people who are going to be in agreement with the recommendation that has been made for the decision.
- Perform – people who are going to put this decision into action.
- Input – people who are going to supply information to those making the choices.
- Decide – people who make the decision and make sure it is carried out.

The GOOD acronym is used in hospitals for communicating with patients and others involved in making difficult decisions. Here's what the letters stand for:

- Goals – the 'big picture' goals for patients, families, and care providers.
- Options – the care options.
- Opinion – remember to separate data from opinion and explain that opinion.
- Document – so that a record of the discussion and decision made exists.

That's a slightly unusual decision process, but clearly it is a decision process.

Each of these acronyms seems to take a similar view of the decision-making process, while at the same time varying in the degree of granularity that it has in certain areas. And each was created for decision-making processes in different areas. Basically the steps seem to be to:

- Identify the need for a decision to be made.
- Look at the choices available, including the option of doing nothing at this time.

- Weigh up the pros and cons of each option – and this can involve looking at the bigger picture and getting input from a number of sources.
- Make a decision.
- Act of that decision.
- Review its impact and see whether the process needs to start again.

ILWMAR doesn't make a great acronym, but I think that it covers the main points.

So, what can you advise your clients to do in order to make a decision about anything? The answer appears to be 'go with their gut' – do whatever feels right at the time, utilize their powerful subconscious mind. There's evidence that people can pick the odd-one-out from 650 identical symbols on a screen better when they were given no time at all than when they had plenty of time. Similarly, an experiment where people had to choose the best car out of four, each with 12 desirable attributes found that, with plenty of time, people were no more accurate than chance (25%). However, when the participants were distracted with puzzles, more than half of them managed to pick the best car.

There is one caveat to this advice and that's to check how emotional they feel about their choice. It can be very easy to go for the selfish option, and then they spend time justifying it to themself. If they do have time to list all the good and bad features of each option, then ask them how they feel about each choice. And don't expect everyone (or anyone, sometimes) to agree with their decision. And if it does turn out to be a bad decision, then you can hope that they will learn from it. Remember that the foundation of wisdom is experience (and that usually comes from making bad decisions).

Robin enjoyed the relaxation and bucket emptying parts of our sessions. He was pleased to find that difficulty with decisions was quite normal and there had been so much research into it. And he quite liked trying the different techniques for decision making. He found that he started with simple choices – such what he wanted to drink at the pub – and built up to what to choose from the menu, and he felt confident that he could make other more difficult choices. He said he felt more relaxed because he wasn't worrying all the time about not being able to make a decision. And he was sleeping better, which helped so many other aspects of his life.

References:

https://www.verywell.com/problems-in-decision-making-2795486

http://projectmanagementhacks.com/the-8-threats-to-effective-decision-making/

http://www.uncommon-knowledge.co.uk/articles/making-decisions.html

https://en.wikipedia.org/wiki/Decision-making

https://www.psychologytoday.com/blog/choke/201104/got-go-wait-youll-do-your-best-thinking

Energy – spinning top

A script to help 'stuck' clients feel more like doing something.

We've all had days, when we wake up and we don't have any enthusiasm for the day. When we get up purely out of habit. And we've all had days where sitting on the sofa seems like a much better option than getting up and doing something – like going to the gym or even just going into the kitchen to get a drink. And we've all sat at work hoping that no more e-mails will arrive with things for us to do.

On the other hand, we've all had days when we've been really buzzing with energy. When we complete one job and go quickly onto the next. Or where we can play a sport or dance for ages and ages. And, at the end, although we feel tired, we feel really good. We're enthusiastic and keen and confident and capable, and it feels really good.

What I want you to do now is remember a time when you felt full of energy, when you felt able to face any challenge. And if you can't recall ever feeling quite like that – that's OK. Because all you have to do is imagine what feeling full of energy, what being ready for anything, would feel like. Just pause for a second while you imagine or remember that feeling throughout your body. Feel the strength, the energy, the power. Contemplate for a moment some of the things you might do when you feel like this.

Pause

And now I want you to imagine that in front of you is another version of you. But this version of you is feeling ten times more powerful, ten times more energetic, ten times stronger. And I want you to step into that version of you and just let yourself feel what it's like when you're ten times stronger, ten times more energetic, ten times more powerful than you've ever been in your life. You're like a racing car at the start of a race, revving your engine, so enthusiastic to get going – so enthusiastic to get all those jobs done and try your hand at all sorts of new things. Let's just stay in this body, feeling those powerful sensations, enjoying the knowledge that you can achieve anything you want to. It's like having some kind of motor inside you, revving away to get started. Such a powerful feeling. And I want you to really immerse yourself in that feeling – enjoy it. Imagine, while you're there, that you can see a spinning top spinning on its axis so quickly.

And now I want you to imagine that in front of you is another version of you. But this version of you is feeling ten times more powerful, ten times more energetic, ten times stronger than the current version of you. And I want you to step into that version of you and just let yourself feel what it's like when you're ten times stronger still, ten times more energetic than before, ten times more powerful than you've ever been in your life. You're like a rocket ship on the launch pad with the countdown almost at zero. Like some amazing children's spinning top that's spinning faster than you've ever seen before. You're so enthusiastic to get going, to escape the Earth's gravity, to start your journey into space – so enthusiastic to get all your jobs done and experience all sorts of new things, not worrying about whether you succeed or fail at first. Knowing that when you learned to ride a bike, you had to fall off a few times before you got the hang of it. And that's true of learning to drive a car, play the piano, and just about everything else. Getting it wrong was just part of getting it right. Let's just stay in this body, let's enjoy those feeling of such power and enthusiasm. It's like such a powerful rocket engine inside you. You just want to get going, to get started on so many things.

And let's take those feelings of energy, enthusiasm, and power with you back to your body. Let's take those ideas of things to do and the keenness to do them back into your own body. Keep the knowledge of those feelings and those ideas with you.

And you can regain those feelings of energy and enthusiasm anytime you want to simply by thinking about that spinning top. And when you picture a spinning top in your mind, you'll feel ten times more enthusiastic, ten times more energetic, than you did before. And those feelings will increase and increase until they reach the rocket-ship-level that you felt here today. And you can use those feelings whenever you want.

And you'll be amazed at how often you'll find yourself getting that feeling of a revving engine or a rocket ship about to take off. As more-and-more you want to make a start on doing things. As you find yourself keen to try new things. And you'll have the energy to do them. And you'll be so pleased that you did take the opportunity to try new things and see what they're like. Just think of a spinning top, and enjoy those strong feelings of energy and enthusiasm.

It's not what it seems

This script can be used to encourage clients to accept that there are things in their life that they can't change, and get on with the good things in their life.

We've all done it, we've all decided that a particular event in our lives is awful, and we've all predicted that if the event were to occur in the future, it would be equally awful – or maybe worse than that. And we've all tried to change things that are completely beyond our power to control.

The Stoics, who spent time thinking about things like this between 3BCE to 3CE, came up with some ideas that are worth considering.

They said that there are no good or bad events, there's only perception. Shakespeare had Hamlet say something similar: "…for there is nothing either good nor bad but thinking makes it so". And psychologist Albert Ellis suggested that next time you're feeling a negative emotion, don't focus on the event that you think "caused" it, ask yourself what belief you hold about that event. And then ask yourself whether it's rational.

Ellis suggested that if you revise your beliefs, you can change your feelings.

Or have you come across Reinhold Nieburh's prayer? The one that goes: "God, grant me the serenity to accept the things I cannot change, Courage to change the things I can, and Wisdom to know the difference". This is also an idea from the Stoics. They would ask themselves the question, "Can I do anything about this?" If they could, then they would. If they couldn't… then they didn't. They said that worrying achieves nothing but stress. And no amount of stressing about something is going to change it. As Erma Bombeck said, "Worry is like a rocking chair: it gives you something to do but never gets you anywhere".

Not only are you going to be happier if you can make the distinction between what you can change and what you can't change, but, if you focus your energy exclusively on what you can change, you're going to be a lot more productive and effective as well.

Next time you're worrying, stop and ask yourself, "Do I have control over this?" If you do have control, stop worrying and get to work changing things.

But if you don't have control, worrying won't make things any better.

And, as Ellis suggested, it might be a good idea to ask yourself what your belief is that's causing all this worry

So if there's no point being sad, angry, or worried, what should you do? Accept it – as simple as that. Ellis, the psychologist, said that people would be happier if they removed words like 'must' and 'should' from their vocabulary. Although you can say, "I would prefer it if…"

So, accept the facts as they are and then decide what you're going to do about them. The Stoics said, "Let's not waste any energy fighting things that are outside our control, let's accept them, let's embrace them, and then let's move on and see what we can do with it."

Epictetus came up with the idea of 'projective visualization'. Suppose, he says, something happens and it makes us angry. One way to avert this anger is to think about how we would feel if the incident had happened to someone else instead. What would we say to comfort them. Engaging in projective visualization, according to Epictetus, will make us appreciate the relative insignificance of the bad things that happen to us and will therefore prevent them from disrupting our natural tranquillity. We might start to live by a rule that says, "Do unto yourself what you would recommend to others".

Or what about that moment – however brief – when you decide whether to give in to an impulse or resist it. You have a choice. All too often, you give in to that habit, you act out that script you've performed a thousand times before, even though it always had unfortunate consequences. The Stoics had a way for dealing with this too. According to Epictetus, the key was that moment when you're deciding. Catch yourself when you're about to act and just postpone it. Just say to the impulse, "Hold on a moment".

It's suggested that telling yourself, "I can have this later" operates in the mind a bit like having it now. It seems to satisfy the urges to some degree – and can be even more effective at suppressing the desire than actually performing the action. It takes much less willpower and it's much less stressful to say 'Later' rather than 'Never'.

And if it's something that happens over and over again, don't try to break the bad habit, replace it with a new one – a new way that you'd like to behave in those circumstances.

Epictetus also wrote: "He is a wise man who does not grieve for the things which he has not, but rejoices for those which he has". You might try going without tea or coffee for three days and see what it tastes like when you drink it on day 4. You appreciate it again, and savour the moment. Focus your attention on it. And focus your attention on other things in your life and really appreciate them. You have so much to be grateful for and yet, perhaps, spend so much time worrying about the things you don't have.

And each evening, review your day. Run through the things to be grateful for.

It's amazing what wisdom we can learn from 2000 years ago if we want to stay calm and in control and really appreciate our life.

Clinic culture

If you set up your own clinic, how do you set the tone of the culture at your clinic?

It might just be that solution-focused hypnotherapists are just solution-focused hypnotherapists. They don't really care which clinic they're working out of on a particular day – they just do solution-focused hypnotherapy. They're experts at solution-focused hypnotherapy – and one clinic is pretty much like another. The only difference is the commute time and the cost of renting a room. But now some therapists are setting up their own clinics and it's time for them to think about what they want the culture of their clinic to be. Different clinics are different (not just whether there are any biscuits). Therapists can tell where they work by the ambience and mood of the place (not just the quality of the coffee). That's because different clinics have different cultures

You hear people say things like: "Culture eats strategy for breakfast", which is attributed to Peter Drucker. And you see people nod their heads as though the statement actually means anything. Of course, 'culture' and 'strategy' are words that you can't put in a wheelbarrow – they don't refer to real things. So what I understand by culture and strategy may be quite different from what you think those words mean. But what it does flag up is that people think culture is important.

So, if we did all think the word culture means the same thing, what would we think it means? The Business Directory (http://www.businessdictionary.com/definition/organizational-culture.html) defines corporate or organizational culture as: "The values and behaviours that contribute to the unique social and psychological environment of an organization". It goes on to say: "Organizational culture includes an organization's expectations, experiences, philosophy, and values that hold it together, and is expressed in its self-image, inner workings, interactions with the outside world, and future expectations. It is based on shared attitudes, beliefs, customs, and written and unwritten rules that have been developed over time and are considered valid. It's shown in:

1   the ways the organization conducts its business, treats its employees, customers, and the wider community;

2   the extent to which freedom is allowed in decision making, developing new ideas, and personal expression;

3   how power and information flow through its hierarchy; and

4   how committed employees are towards collective objectives.

"It affects the organization's productivity and performance, and provides guidelines on customer care and service, product quality and safety, attendance and punctuality, and concern for the environment.

"It also extends to production-methods, marketing and advertising practices, and to new product creation. Organizational culture is unique for every organization and one of the hardest things to change."

People like Schein (1992), Deal and Kennedy (2000), and Kotter (1992) have complicated things by suggesting that organizations often have subcultures. I guess it depends how large a clinic you're setting up whether the acupuncturists group together and create their own subculture!. The authors suggest that these co-existing or conflicting subcultures exist because each subculture is linked to a different management team. For smaller clinics, this shouldn't be an issue so long as each therapist working in the clinic feels part of the whole clinic and not isolated in any way.

James L Heskett has suggested that culture "can account for 20-30% of the differential in corporate performance when compared with 'culturally unremarkable' competitors". So which companies have outstanding cultures? Google is usually held up as an example. It provides free meals, employee trips and parties, financial bonuses, open presentations by high-level executives, gyms, a dog-friendly environment, etc. In return, its staff are driven, talented, and among the best of the best. Similarly, Facebook provides lots of food, stock options, open office space, on-site laundry, a focus on teamwork and open communication, a competitive atmosphere that fosters personal growth and learning, and great benefits. On the down side, a free and organic organizational structure that worked for the smaller organization is less successful for the larger one. Adobe likes to avoid micromanaging its staff, and managers take on the role of a coach, more than anything, letting employees set goals and determine how they should be assessed. Employees are given stock options and continual training. Thanks to Sujan Patel at https://www.entrepreneur.com/article/249174 for this information about corporate culture.

How does Google's, Facebook's, or Adobe's culture compare to your new clinic? I wonder. So, what are the components of great culture? John Coleman in an article at Harvard Business Review (https://hbr.org/2013/05/six-components-of-culture) has some suggestions. These are:

- Vision – a great culture starts with a vision or mission statement. That purpose, in turn, orients every decision employees make. A vision statement is a simple but foundational element of culture.

- Values – a company's values are the core of its culture. Values offer a set of guidelines on the behaviours and mindsets needed to achieve that vision.

- Practices – values are of little importance unless they are enshrined in a company's practices. Whatever an organization's values, they must be reinforced in review criteria and promotion policies, and baked into the operating principles of daily life in the firm.

- People – no company can build a coherent culture without people who either share its core values or possess the willingness and ability to embrace those values. That's why the greatest firms in the world also have some of the most stringent recruiting policies. People stick with cultures they like, and bringing on the right 'culture carriers' reinforces the culture an organization already has.

- Narrative – any organization has a unique history, and the ability to unearth that history and craft it into a narrative is a core element of culture creation. These narratives are more powerful when identified, shaped, and retold as a part of a firm's ongoing culture.

- Place – place shapes culture. Open architecture is more conducive to certain office behaviours, like collaboration. Place – whether geography, architecture, or aesthetic design – impacts the values and behaviours of people in a workplace.

Coleman suggests that there are other factors that influence culture, but these six components can provide a firm foundation for shaping a new organization's culture. And identifying and understanding them more fully in an existing organization can be the first step to revitalizing or reshaping culture in a company looking for change.

This list, perhaps, gives us a better place to start when thinking about creating our own culture in our new clinic. It's important to turn your vision into a vision statement, which therapists working with you can feel is the backbone of how they work. And, although it can often be hard to verbalize them and write them in a list, identifying your core values can help therapists working in your clinic know how you expect them to behave in terms of punctuality and simply interacting with clients and other therapists. Once the values are clearly identified, then people can ensure their practice matches up to them. If they don't, then you, as the clinic owner, need to have a conversation with the therapist. In fact, when you accept therapists to work in your clinic, one of the important interview questions is to check that their own values align with yours. That should avoid any issues later on. And this also meets the 'people' component of great culture. One thing many companies forget to do is to tell their own story. Create a narrative about how the clinic came about and high and low points (in a solution-focused way) in its history. It has a really cohesive effect if everyone is telling the same story. If your clinic is open and sunny, that will have an effect on the mood of the people working there and the clients who attend. Obviously, it doesn't want to be too bright for hypnotherapy. If the building is naturally darker (north facing with smaller corridors, etc) then the decoration is important to brighten things up. Lots of pictures on the wall and bright cushions or bean bags around the place.

It's also important to remember Maslow's hierarchy of needs and make sure the place is warm, but not too hot, and there are coffee and tea-making facilities available (as well as biscuits and sometimes cake) to keep staff happy. And clean toilets for staff and the public to use.

It's interesting that something so ethereal as 'culture' can make so much difference to a clinic. I think it's important that people feel welcome when they enter – whether as a therapist or client. And I think the clinic manager can make a tremendous difference to the mood and tone of the place. But I do wonder, whether some therapists still can't tell which clinic they're in because they all feel much the same. And that's why, I think, if you are setting up a clinic, it's important to get the culture right from the beginning. And to start thinking about culture before you start anything else.

References:

https://hbr.org/2013/05/six-components-of-culture

https://www.entrepreneur.com/article/249174

https://en.wikipedia.org/wiki/Organizational_culture

Working with children

This article takes a look at some of the things you need to know when working with children.

Let's start with the regulations. If you work with children (or vulnerable adults) you need an enhanced DBS check. The AfSFH wants you to renew this every three years. In terms of organizations that you may come across, the Professional Standards Authority (PSA) oversees regulators such as the CNHC (Complementary and Natural Healthcare Council), which you are probably a member of. Professional Associations (like the AfSFH) make sure that their members are appropriately qualified. The Department of Health recommends that people use CNHC members, and the General Medical Council says doctors can refer patients to people who are on the CNHC register. So, it's worth making sure that you are a member. The most recent Act of Parliament concerning the safeguarding of children is the Protection of Freedoms Act (2012).

The CNHC's code of conduct and ethics has some guidance about seeing children. If a child is under 16, another person (usually a parent) should always be present. Confidentiality is central to the relationship between registrants and clients. A parent must give express consent before we can treat their child and this must be in the form of written consent. In terms of the records you keep, both the child and the parent (or a person with parental responsibility) can see them. If there is a safeguarding issue that comes up, this must be reported to Social Services. You may want to think about 'run off' professional indemnity insurance, which covers you when your company ceases trading or you stop practising hypnotherapy, in case anyone decides to sue long after their treatment was over. In addition, if a child is currently undergoing psychiatric treatment, they can only be treated with the doctor's consent.

The age of criminal responsibility (in the UK) is 10. The age an adolescent become an adult by law is 18.

Gillick competence is used in UK medical law to decide whether a child (under 16) is able to consent to their own medical treatment, without the need for parental permission or knowledge.

There are many child development theories proposed by people like Piaget, Erikson, Vygotsky, and Chomsky, but they tend to agree that children are better able to deal with things if they have the right support. Behavioural psychology look at everything in terms of rewards and punishment. It's clear that there are lots of influences on children's behaviour, it's not just parents. These influences include schools, clubs, friends, TV, etc. As children grow up, they go through different stages. The idea of scaffolding suggests that children move across these stages with the help of people around them. Attachment theory suggests that when a parent goes out of a room and returns later, the child may carry on as normal, or try to punish the adult for leaving them, or not re-establish contact at all. It's useful to have some knowledge of these various theories when working with children.

The Child and Adolescent Mental Health Services (CAMHS) are the NHS services that assess and treat young people with emotional, behavioural, or mental health difficulties. They work with young people with depression, problems with food, self-harm, abuse, violence or anger, bipolar, schizophrenia, anxiety, and other issues.

Parents, teachers, and GPs can refer youngsters for an assessment with CAMHS. Various sources suggest that around 40 percent of clients you'll see will have seen CAMHS. And nearly 30 percent of referrals to CAMHS aren't treated because they aren't thought to be serious enough. And, again, may come to you.

Parents often want to talk to the therapist before a session to find out what will happen during the session. They may have already tried other therapies, so show them that hypnotherapy is not just another treatment, it's an appropriate, effective, and simple remedy for their child's problem. Illustrate this with examples of successful treatments. Explain that there are no side effects. And explain that when children are anxious, parents are too. And this feeds back to the child.

It's important that the parent (or the referrer) understands the nature and/or outcomes from hypnotherapy. There are circumstances, where you won't want to take their child as a client, for example if they:

- Are coming for fun
- Have diagnosed/undiagnosed learning difficulties
- Don't want to attend
- Are trying to mask the pain from a sport injury and want to continue training, which may aggravate an existing condition, create a new one, or endanger the child.

You should take a child as a client if:

- Their problem has been shown to be treatable through hypnotherapy.
- The child has some motivation to remedy the problem.
- The parents approve/support the treatment plan.

Always ask about a medical diagnosis/treatment or whether the child is undergoing psychological/psychiatric treatment. If they are, you will need to let them know that you are also treating the child. You should not treat self-harm without the express permission of a psychologist/psychiatrist.

There are usually fewer therapy sessions needed for positive change to occur in children. When working with children there are a number of things to bear in mind. The child is forming their model of the world, so it won't be the same as a mature adult's. So work with the child's model of the world. Younger children (under 12) don't get metaphors – they don't make the connection and see how they apply to them. They just see it as a story. Remember that when you talk to a child, they cannot *not* respond. Respect everything that a child says to you (no matter how bizarre). As with an adult, a child already has all the resources they need. And always show the child that they have a choice about how they act and react. Don't take away their choice. And children usually make the best choices for themselves at any given time.

As with all clients, that Initial Consultation is an opportunity to build rapport with the child. The first questions you might want to ask them are their name, address, and age. They will usually know this, but it's a way of getting to know them (and their competence level). Whatever age they say, say that you thought they were two years older. Some other useful questions are:

- Do they have a best friend/girlfriend/boyfriend?
- Do they have brothers/sisters? How do they get on?
- Ask them to describe their life in one word.
- What time does their mum/dad get back from work?
- Are their parents divorced/separated/with new partners?
- What are their parents' names, contact e-mail?
- Ask about grandparents – their closeness/health.
- Do they like school?
- What are their favourite subjects?
- How are they doing?
- Do they experience bullying?
- Do they have pets?
- How do they spend their day/weekends?
- How do they spend their holidays?
- What are their hobbies/interests?
- Who are their heroes?
- What's their favourite book/favourite films/favourite computer game?
- How long do they spend on computers?
- Do they have anger issues?
- Do they have any frustrations?
- Do they have feelings of guilt?
- Does the child understand hypnosis?
- The Miracle Question
- "If you were prepared to change, what would it take? Eg an iPad, trip to Disney World."

It's important with any client to make sure that they understand what you're saying to them. It's unusual for them to ask. You can do this by asking them to repeat back what you've just told them. The age of a child isn't always a good indicator of their vocabulary. If you look at the child's face while you're talking, you can usually tell whether or not they understand what you're saying. It can be a good idea to summarize what you'll be saying in trance later to make sure that they understand the language that you'll be using. A child's understanding will depend on their age.

Of course, children aren't as good as adults at concentrating. So, for 5-year-olds, 10 minutes in trance is about as long as they can manage. If you need a longer time, use two trance sessions – bringing the child out of hypnosis and re-inducing the state. That means that a session with a younger child may last only 30 minutes at the most – so arrange your time appropriately. Children also tend to accept any statements that you make (unlike adults).

Five is probably the youngest age to work with children. Once they are over 10, it's possible to treat them in a more adult way. So how would you treat youngsters? Involve the parents even though they are often the problem and they often create unnecessary expectation or pressure for change. It's important to be aware that simply seeing you can create an 'excuse' for change to take place. And, bear in mind that children may have invested a lot of time in their behaviour. And suggest that change will take place, but don't put the child under any pressure as to when that change will occur. Find out about the problem as the child sees it. Also, find out how the problem manifests itself physically. And investigate what makes it better and what makes it worse. Also, find out what the child wants from the therapy. The child must want to make the change. The problem that's brought them to see you may be their defensive strategy.

Most sessions with children follow a standard format. Where appropriate, normalize whatever their problem is (lots of people have it/do it). Help them to understand body/mind control, and help them to understand mind/body control. Explain that the child is not responsible for their problem (eg blushing). Although, with anger issues or bedwetting, it is their problem and you need to reinforce their control/responsibility. Explain that their subconscious mind is going to make positive changes to their behaviour. Also tell them that there is no pressure for when the change will happen. The most common type of issue with children is psychosomatic. Again, it's worth summarizing the scripts before being delivered in trance. Note that anorexia is the health issue with the highest mortality rate in girls. It is important to explain the physical symptoms of anxiety to children.

Let's suppose you saw a child with a pain and the doctors said there was no physical reason for the pain. You might suggest that it seems as if muscles have some kind of memory. Your muscles remember how to ride a bike, and you can't forget how to ride a bike – if you did, you'd get on a bike and fall off. And sometimes, it's the same with pain. Your muscles keep remembering the pain that they once felt. The pain has to travel from the muscle to the brain. You can treat this by asking the child to visualize filling the area with a nice colourful liquid (the child chooses the temperature, colour, etc). Ask the child which combinations would make the pain worse and which would make it better. When the liquid drains away, the pain goes. You can say: "And I don't know whether it will be today, or tomorrow, or even a couple of days' time, but I do know that you'll be so pleased when you realize that the pain has gone". Also, tell the parents not to keep asking whether child still has the pain, they should wait for child to tell you them that the pain has gone. If the child says that the pain has worsened, tell them that that's what's meant to happen as the first stage of it going completely.

If a child comes with blushing issues, talk about how the mind and body are linked, and also discuss thoughts and feelings. Talk to them about the brain and the connection between the mind and body, and thoughts and feelings. And explain how things happen without our permission. Ask whether they've ever watched a scary film? How did they feel? They didn't choose to feel scared. Their brain has made a mistake, because, in reality, they are perfectly safe sitting next to mum and dad on the sofa. They just had a feeling. Or, if they used to believe in Father Christmas, how did they feel? They probably got excited when they thought he was coming. Or what happens when they are sad? They begin to cry. They don't decide to cry, it just happens. Explain that things happen in their body without them wanting them to. Or it may have been appropriate once, but it's not now.

You may see children with ADHD (Attention Deficit Hyperactivity Disorder). They are often impulsive, fidgety, and talkative. The best way to treat them is to focus on the impulsivity. Impulsivity may depend on the child's interests. They tend to be more impulsive when they are less interested. Your sessions will focus on control, using the strengths they have to stay focused in the areas they have an interest and use those strengths in areas where they are less interested.

You can work with children with conversion disorder. This is a psychosomatic disorder, such as numbness, blindness, paralysis, or fits, which doesn't have an organic cause, and can usually be traced back to a traumatic event.

You may see children who feel guilty about something. Explain that guilt is a feeling. It doesn't need to have any facts behind it. It may not be their fault, it's just how they feel. Ask them whether there are any reasons that they should feel that way, and explain that feelings and thoughts are different things. And help them to move on. If they are guilty for what happened, try to understand the circumstances in which the event happened. Separate the behaviour from the person. It may have been an accident, they may not have meant the consequences to occur.

With bedwetters, it depends on their age. If they are old enough to have dry nights, ask whether they have seen a doctor. There may be a medical reason. Usually, the child will have been told that they sleep too deeply that's why they wet the bed. You need to give them a different model. Focus on dryness. Explain that their brain is working while they sleep. Their bladder sends message to the brain saying they need to go to the toilet. Their brain then has a choice (same as during the day). It can choose to get them up and go to the loo or not. Currently, it has the wrong habit. At the moment it just says 'fine', and the child wets the bed. So focus on establishing a new habit at night time.

Encopresis is the medical term for a toilet-trained child (aged four or older) soiling their clothes. The soiling usually happens during the day, when the child is awake and active. Often this is unconscious, although sometimes it's because the child likes the sensation. Again, it's a case of creating new habits.

Nightmares may come with a secondary gain in that the child gets to sleep in mum's bed. Often the child thinks they can see things in the semi-darkness of their room. Get the child to tell you what they think they can see in detail. Often, large chunks are missing. You can change the picture, eg add a friendly teddy bear's face to fill in a dark shape. The child often comes up with ideas of what can go in blank areas. Also, look at their sleep pattern. Look at what happens before they go to sleep. You can help with what they are thinking about before they go to bed.

If a child has a fear of dogs, this may be sensible because some dogs are very big when compared to small children, and some do bite. So children do need to be cautious with dogs. One technique you can use to help with their fear is the hero technique. David Beckham, for example, is walking his well-behaved dog. He lets the child stroke his dog. And then he lets the child take the dog for a walk. Beckham then says what a brave little boy/girl the child is.

You may see children who only eat a limited diet. The first thing to establish is whether the child wants to change. Has a doctor said he's OK on that diet? There may be difficulties around smell, texture, or taste. And some foods are harder to swallow than others. You can get the child to agree how many more things they want to be eating by the end of the month. Get them to agree to try different foods, and agree (with the parents) that they can do it when no-one is watching them.

Anger issues are common. Find something they can control – often they are restricting their own behaviour, preventing it being too extreme. You need to find where they draw the line. And then help them move the line. Ask them what's going to happen if they carry on behaving the way they are. Also ask if they stop, what's going to happen? (This is the Dickens Process.) Ask them whether carrying on with their behaviour fits in with their long-term goals. Is it a good idea?

If they have self-confidence issues, find out what they can be confident about. Problems around social comparison are usually caused by social media.

If they have problems with bullying, attempt to disenfranchise the bully. Make the child think the bully looks weak. Ask why the bully behaves the way they do. Do everything to diminish the bully.

Cyberbullying is all to do with what other people are thinking. Ask the child how do they know what other people are thinking? The answer will be because they've just said so on Facebook (or whatever). Just because someone says/writes something doesn't mean that's what they're really thinking. Often they are just saying something to get a reaction from you. The child thinks everyone else is thinking what they're thinking. The question is, is everyone thinking the same thing? Of course, what other people think is to do with them not with you. It's based on their experience, not yours.

As mentioned above, don't work with children who self-harm unless it's agreed with their psychiatrist. Cutting is a form of release. They can also get that feeling of release by breathing out. Self-harm has the greatest risk of suicide in children and adolescents.

If a child bites their nails (and their fingers), find out whether they know they're doing it. Suggest that it's automatic, although there were reasons why they started doing it. If they do it 10 times a day, that's 3.5 thousand times a year. That's a lot of investment in the habit. Build on the idea of nice nails. And build on positive outcomes.

Children may come with sport performance issues. They are often fine when they are being reactive, but when being proactive (ie thinking about what they are going to do) they struggle (eg when serving in tennis or taking a penalty). Sometimes they have to deal with supporters of their opponent being rude, which can affect their confidence. They may also find that they can only focus for shorter periods of time. Top players stay in focus longer. Helping with confidence can help them. But don't help them ignore the pain coming from an injury. They need to repair the damage and then carry on with their sport – no matter how important the competition coming up is.

If a child comes who stammers, check whether they've had a medical examination or are having speech therapy. Stammering is usually anxiety-related. Find out when it doesn't happen. Try getting the child to talk in trance. They usually can without stuttering. Find out in which situations it happens (it may be just some words or all words). Get them to talk when breathing out. Focus on breathing and not on words.

A child who sucks their thumb does it unconsciously. However, they may notice that it's pushing out their front teeth and they have a bite mark on their thumb. It's also very noticeable to their friends. Like hairpulling (trichotillomania), using a mirror so they can see themselves doing it can be very effective.

With weight issues, the first thing to find out is what the parents are feeding the child. What would be the outcome if they carry on? After that, it's sensible portion sizes, sensible eating, and exercise.

Eczema is anxiety-related. Children will scratch their skin at night. You can get the child to imagine a soothing lotion on their skin – perhaps a guardian angel administers the lotion while they sleep. Or they can imagine a beam of light shining on them while they sleep making them better.

Emetophobia can often result from how people reacted the first time they were sick. Treat in the usual way – bucket emptying and confidence building.

Fainting is lack of blood to the brain. The reasons could be physical, anxiety, or sudden shock. Some children are fearful of it happening again – perhaps it happened in public and so they were embarrassed. Explain to the child why it happened and normalize it. Children usually get over it quite quickly. Tell them it probably won't happen again.

If a child has hiccoughs, stare into their eyes and suggest that they close their eyes. While you're doing that, they won't hiccough.

If a child is worried about choking they can be stressing the muscles in their throat, which can stop the tongue moving properly. They need to relax the muscles of their throat, then they'll be able to swallow properly. They may have the feeling because choking happened to them once and they are worried about it happening again.

Children with selective mutism are meant to be trying to talk (in theory), but often it's more convenient for them not to talk.

If you see a child with chronic fatigue syndrome, try to normalize it and take away their responsibility. (A syndrome is a cluster of symptoms that occur together without a known cause.)

Most of the other symptoms you'll see children with can be treated in similar ways to those listed above.

References:

https://www.tisph.com/

Lynda Hudson. Scripts and Strategies in Hypnotherapy with Children. February 2009. 978-1845901394

Adding play therapy to your hypnotherapy skills

A review of what play therapy has to offer hypnotherapists.

More-and-more hypnotherapists are seeing younger clients, and I wondered whether it would be useful to incorporate some aspects of play therapy into the sessions to help these youngsters overcome whatever issue it was that brought them to see us.

According to a study in 1993 by Charles Schaefer, play is equally important for human happiness as work and love. Another study by Garry Landreth in 2002 found that play is not only an enjoyable and fun activity that stimulates both the human mind and body, but also something that improves people's outlook on life. It also aids increasing self-expression, self-realization, and self-efficacy. A similar study found that play helps increase a person's ability to learn and develop skills.

Play therapy can help children learn how to express themselves in a manner that is constructive rather than destructive for their personal relations and themselves. Play therapy encourages strong decision-making skills as well as acceptance of responsibility. It's an excellent way to help with the treatment or recovery of a child from a past traumatic or stressful experience. It's very effective with children who have low self-esteem or aggression issues.

So, play is a major stress reliever that can help with emotional stability and provide an ego boost. It also teaches useful social skills and helps people to cope better with difficult situations. And it helps children to resolve inner conflicts by helping them to learn adaptive behaviours if they lack social or emotional skills. Play therapy is the only tool that allows children to express their inner feelings because toys become their world and play is the language they use to express themselves. It also allows children, who have suffered mental trauma, to be able to communicate better, express their feelings more easily, and develop skills for problem-solving.

History and leading lights

Play Therapy was developed in the twentieth century, although the idea of play as therapy has been alluded to since the time of Plato around 400 BCE. Rousseau wrote about the importance of play in his book, *Emile* in 1762, as did Friedrich Frobel in *The Education of Man* in 1903. Play therapy was first used by Sigmund Freud in 1909 on Little Hans (a 5-year-old child with a phobia).

The process of therapy was formalized by Hermine Hug-Hellmuth in 1921, who provided children with various playing materials so that they could express their thoughts and feelings. Melanie Klein began using this technique in 1955. Anna Freud also made use of play therapy.

David Levy developed release therapy in the 1930s, where a child who is experiencing a certain stressful situation plays freely. The child can deal with the trauma by using toys to relive it. Levy's work was then taken further by Gove Hambidge in 1955, by emphasizing a structured play therapy model. Relationship therapy was developed by Frederick Allen and Jesse Taft.

In 1942, Carl Rogers developed a technique that was somewhat non-directive in nature, which was later known as client-centred therapy. Virginia Axline, who was mentored by Rogers, worked further on these concepts. She wrote an article that was called *Entering the child's world via play experiences*, in which she spoke about play therapy in great detail.

Clark Moustakas wrote a book in 1953 called *Children in Play Therapy*, and went on to compile the *Publication of the Self* in 1956, which was the end result of a conversation that took place between himself, Carl Rogers, and Abraham Maslow, among others. This eventually led to the formation of the Humanistic Psychology movement.

In the 1960s, Bernard and Louise Guerney developed filial therapy. This focused on a very structured and organized training program that was meant for parents. The purpose of this training program was to tell them how they could also have child-centred playing sessions while at home.

The 1960s saw an increase in the number of school counsellors, increasing the number of people making use of play therapy. The counsellors began using the play therapy approach as a preventative tool and also for educational reasons in order to deal with the many issues that children were presenting with.

Clark Moustakas published another book on play therapy that was called *The Child's Discovery of Himself*. In this book, he talked about therapy as if it were a growth experience.

Well-known play therapists in the USA are Landreth, Moustakas, and Schaefer who developed their own slightly different models, which brought in various other elements of therapy, such as systemic family therapy, cognitive behavioural therapy, and solution-focused therapy.

In the UK in the 1980s, The Children's Hour Trust began training using the model that was developed by Virginia Axline. A British play therapy movement was initiated by Ann Cattanach and Sue Jennings, and, in 1990, the Institute of Dramatherapy also began offering a Diploma and Certificate in play therapy.

The British Association of Play Therapists started in 1992 by people studying the approach at the Institute of Dramatherapy. This led to the British Play Therapy movement.

Play therapy models

There are a number of play therapies that can be tried.

Sensory and Embodiment play therapy uses the child development theory of Embodiment Projection Role (EPR), which studies the development of children from birth to the age of seven through drama, and is also referred to as dramatic development. This theory was created by Dr Sue Jennings. Embodiment play involves several techniques such as:

> Play therapy helps children understand muddled feelings and upsetting events. Rather than having to explain what is troubling them, children use play to communicate at their own level and pace – without feeling interrogated or threatened.

- Body movement as a whole
- Fine body movement where the client only moves a specific body part
- Dancing
- Sensory movement involving the five senses: touch, sight, smell, hearing, and taste
- Physically exerting games, such as wrestling.

Some of the techniques used in Projection play therapy are:

- Playing with substances such as plasticine, sand, water, etc
- Using pictures or drawings to project
- Playing with toys
- Creating scenes with dolls, puppets, etc
- Playing with natural objects such as leaves, pebbles, and twigs.

In the Role stage, children get to be involved in dramatic play. Some of the techniques in this therapy are:

- Creating characters based on animals that can speak
- Making stories a basis for the scenes that will be enacted
- Using masks to help with the role playing in the story
- Writing a TV script and acting it out together.

Dramatic Role Playing can involve creative dramatic play, role play, child-toy interactions, and play acting.

Art therapy uses creative media of self-expression along with psychotherapeutic techniques to resolve deep-rooted psychological problems. It's used as a way for children to communicate their problems, which, otherwise, they would find hard to do. The artistic methods that can be employed by the therapists range from drawing, painting, and sculpting to collage making.

Art in play therapy can help to resolve issues such as:

- Mental health problems
- Grief over loss of a loved one
- Physical or sexual abuse
- Life threatening diseases, such as cancer
- Schizophrenia in children
- Emotional or cognitive disabilities.

Sand pits and creative mosaics for play therapy allow a child to express themselves without having to use words. This is particularly useful for play therapy because children find it extremely difficult to use words to describe what they are feeling. When given a medium of art, children are easily able to describe their innermost emotions and thoughts.

Storytelling can be a strong tool in play therapy. It is very useful as a developmental tool because stories elicit children's thoughts and feelings, and allow them to better understand their world. In addition, the world of stories is familiar to children, and it is easy to engage them in this type of therapy.

The mutual storytelling technique was developed by Richard Gardner in 1972. In this technique, the child is first asked to tell a story; it has to be one the child hasn't heard before, and should have a beginning, a middle, and an end. The therapist listens to the story and analyses the psychodynamic meaning of the story. When the child finishes telling the story, the therapist tells a story of their own; it has the same characters and setting, but is a healthier adaptation and has better conflict resolutions. While the child is telling the story, the therapist can ask questions to clarify certain aspects. The story the therapist tells should have a moral that can be used to further the child's understanding of emotions, empathy, and relationships. This technique is especially useful in cognitive behavioural therapy for children.

Modelling storytelling involves identifying a specific skill that the child needs to be taught, and then developing a story focusing around that skill. There are nine major skills:

- Closeness and relationship building
- Handling separation and independence
- Handling decision making and internal conflicts
- Dealing with frustration
- Celebrating good things and feeling pleasure
- Delayed gratification
- Relaxing
- Cognitive processing
- A sense of direction.

The technique involves displaying a collection of random pictures in front of the child and the therapist. The pictures should represent the new skill that is to be learned. The therapist picks a picture first and develops a story about that skill. After the therapist, it is the child's turn to pick a picture. The stories being told should have a beginning, a middle, and an end, and should also contain a moral.

Three techniques that can be used to treat low self-esteem are:

- Pretend play – this allows the child to project their own difficulties onto a toy or puppet. The child can be asked to put on a show that focuses on the puppet going through the same problems that the child is. And also ask the child to come up with solutions for the puppet's problems.
- Encourage independence – when a child refuses to do something or asks an adult to do it, gently persuade them to do it on their own. Tell them that you believe they can do it, and keep encouraging them to carry out the task.

- Self-awareness – encourage them to answer questions such as who they are, what things they like, what they like to eat, etc, and be supportive of their answers. Help the child to accept their answers. If they are finding it difficult to give an answer, give them a choice between two options and ask them to choose either one.

Play therapy techniques

Ways to improve a child's self-esteem include:

The 'feeling' word game was developed by Heidi Kaduson. This involves the therapist asking the child to describe feelings a child their age would have. Each feeling is written on a different piece of paper and the papers are lined up in front of the child. The therapist then tells a story about themselves, including positive and negative emotions. As each feeling is described, a poker chip is put on the paper with the relevant feeling written on it. These chips are then put into a tin at the end of the story. After the therapist has finished, they should encourage the child to tell a story and put poker chips on the paper as they go through different feelings. This play can continue the same way until the present problem is discussed.

Colour your life was developed by K J O'Connor in 1983. This technique allows children to understand and discuss their various emotional states in a non-threatening manner. The therapist starts by asking the child to make various colour-feeling pairings. For each colour, the child is asked to denote a feeling. For example, red denotes anger, blue denotes sadness, until most emotions have an associated colour. Next, the child is given a blank piece of paper along with some crayons to colour the paper according to what emotions they have felt in the past. They can draw in whichever way they feel like, in geometric shapes or designs. If there needs to be a focus on recent life events, then the child can be asked to draw their feelings about that certain event.

Balloons of anger was developed by Tammy Horn. This teaches children how to release their anger appropriately. Balloon of anger provides children with a visual representation of what anger is and what it does to them if they do not release it slowly and safely. It shows them how anger can build up and what sort of an impact that has on their environment. The technique involves the child blowing up a balloon, which the therapist ties up. The therapist then explains to the child that the balloon represents the human body and the air is anger. The therapist helps the child understand the analogy by asking questions such as, "Can the air get out of the balloon?", "What would happen to you if this anger was stuck inside you?" etc. Then the therapist asks the child to jump on the balloon till it explodes and all the air or anger comes out. The therapist then explains what would have happened if this balloon were a person and how the air coming out signifies an aggressive act, which would hurt the people around that person.

The next step is to have the child blow up another balloon but instead of tying a knot, the child is asked to keep the end pinched closed. The therapist asks the child to release some air and then close the end again. While repeating this the therapist can ask questions such as "Did the balloon get smaller?", "Did it explode?", "Did the people around the balloon get hurt?" etc. At the end, the therapist again explains how the air represented anger and that it is more appropriate to release it slowly and safely. After the activity is complete, the therapist can talk to the child about anger management.

The mad game was developed by Patricia Davidson. It shows children that anger is an acceptable emotion to feel, and allows them to express it verbally. The therapist takes plastic blocks and distributes them between themself and the child. Turn by turn, the therapist and the child stack blocks on top of one another while expressing what makes them angry. It can be anything from something silly to serious. The therapist can start with benign issues first and eventually move on to more specific issues that the child faces. When the blocks have all been stacked, the child is asked to express one thing that makes them very angry, make an 'angry face', and topple all the blocks down. Anger could be replaced by sadness or anxiety.

Beat the clock was designed by Heidi Kaduson. This technique increases a child's self-control and impulse-control. The goal of the game is for the child to remain focused on the task at hand and not get distracted for specific amount of time. The game starts with the therapist giving the child some blocks and ten poker chips. The child is instructed that they have to build the blocks in a specific amount of time; a timer is set for ten minutes. The child is also told that they need to only concentrate on the blocks for ten minutes. If the child looks away from the blocks, then they will have to give the therapist one chip. If they are able to complete the task without getting distracted, they get to keep all the poker chips. Once the child has collected 50 chips, they can choose a toy from the treasure box. This acts as a motivation for the child to concentrate only on the task they have been given. It is very useful for children who have impulse-control problems or concentration issues.

Worry can was designed by Debbie Jones. This helps children to express things that worry them. Children tend to keep those things bottled up inside and this leads to problems in their lives. This technique helps these children to identify and discuss their worries with their therapist. The therapist starts by cutting a piece of paper large enough to cover a tin can. The child is then asked to either draw or write scary things on the paper and colour it with felt tips. The child then glues the paper onto the can and the lid is placed on the can. A slot is cut on the lid of the can, large enough to let small papers slide through. Next, the child is asked to write their worries on separate pieces of pieces of paper and put them into the can. The child should share some of their worries with the therapist while they are putting the papers in the can. While talking in the third person (to allow the child to distance themselves from the problem) the child is asked to provide solutions to the things that worry them. If the child is unable to do so, then the therapist can make suggestions while still maintaining the pretext of play.

Clearly, any hypnotherapist who wants to use these techniques all the time with children needs to complete a play therapy training course. However, some of the ideas and techniques used by play therapists could be incorporated into the first half of a hypnotherapy session to help the child start to deal with whatever issue brought them to see us, and so get the most benefit in the shortest time from our hypnotherapy techniques.

References:

https://en.wikipedia.org/wiki/Play_therapy

http://www.coe-onlinetrainingcourses.com/a-z-of-courses-c4/play-therapy-course-p187

http://www.withkids.org.uk/what-we-do/play-therapy

Pushing the envelope

Here we take a look at how the language and techniques used in hypnotherapy can be used in politics and other areas.

One of the things we, as hypnotherapists, do is tell our clients to make changes in their lives. This probably isn't done directly, but in a more Ericksonian way. We use metaphors, and other imaginative experiences. There are many books and articles suggesting that this way of talking can be extended into everyday conversation resulting in the subtle manipulation of the person being spoken to. This may or may not be possible. Just because you can read about it on the Internet doesn't necessarily make it true. But this idea of conversational manipulation can be extended to public speakers (usually politicians) who want to win round the emotions of a crowd and get them on the speaker's side. Again you might argue that the crowd aren't being hypnotized and it's completely different from the sorts of technique that we use. Nevertheless, I thought it would be interesting to look at some of the techniques that public speakers use to manipulate – dare I say, mesmerize – a crowd.

Politicians seem able to play on the emotions of people. A bit like football teams, discussions about politics can often move from the logical to the emotional very quickly. For a start, politicians are going to talk about things that you can't put in a wheelbarrow. They're going to use words that have a slightly different meaning when I use them to when you use them. And those words come with emotional baggage. They're going to use words like 'immigration', 'freedom', 'sovereignty', etc. They are then going to use metaphors – again a way of talking that has a different meaning to each listener. Metaphors can help to make an abstract idea more concrete. And, again, those metaphors are going to be emotionally charged. They'll talk about "taking back control", "hardworking families", "making Britain great", and other similar phrases.

Another technique used by effective public speakers is repetition. Think of New Labour's slogan in 1997 of "things can only get better". After enough repetitions, you find yourself saying it! And if you can repeat an idea just often enough for people to remember the phrase but not so often it's permanently associated with you, they might even think that they thought of it.

Thinking about well-known public figures, it's quite easy to feel quite emotional about them – in the sense that you either love them or hate them. Many politicians will try to connect to you at an emotional level. They will try to get you very upset or angry about some state of affairs (perhaps the number of foreigners taking our jobs or some other hot topic for the audience they are speaking to framed using strongly negative words) before they explain how voting for them will result in this travesty of the natural order of things being brought to an end (from their point of view). You can think of this as anchoring. They will anchor positive feelings about themselves and anchor negative feelings about their political opponents.

Politicians will also employ hypnotic themes. Hypnotic themes are like powerfully emotionally words, which paint a much bigger and fuller picture of the image that the speaker wants the crowd to take a way with them. The speaker might talk about freedom and then use the word to illustrate different aspects of its meaning in different contexts. This can be very persuasive.

And politicians will pepper their speeches with power words. Power words in this context are words that already have an emotional content – words like safety (as in safe in your own homes), rights (as in your right to do something), drugs (as in the war on drugs – which is a metaphor in itself). Power words are like emotional shorthand. They immediately paint a mental picture in the mind of the person listening. They get the listener's imagination working. Public speakers will often use the power words that we might use in a trance situation, words like:

- Imagine
- Just pretend
- Suppose
- As
- The more
- Every time
- Because
- Which means
- And
- What's it like when
- What would it be like if
- Remember
- Find yourself
- Realize
- Sooner or later.

Are they hypnotic techniques? I'm not sure. A mob is not too dissimilar to an emotional crowd that has been whipped up by a powerful speaker. Both groups of people seem to be bypassing the executive control functions of their brain – their critical faculties.

Whatever it is that's going on, it's quite interesting to see how similar it is to the techniques that we use for hypnotherapy and see whether there is anything we can learn from it to make our therapy sessions more effective.

References:

https://hypnosistrainingacademy.com/3-surefire-power-words-to-gain-power-and-influence-people-fast/

http://www.adam-eason.com/politicians-really-using-hypnosis-us/

https://hypnosistrainingacademy.com/hypnosis-in-politics-hypnotic-language-techniques/

The good, the bad, and the can't be bothered

A sideways look at how above-average hypnotherapists can help clients actually do something.

I was going to write an article about dealing with procrastination – either your own or the procrastination of your clients – and I probably will one day! For the moment, I just need to see what all the people that I'm friends with on Facebook (but never actually meet) have been doing recently. And then I'd better check what the latest news is, in case anything good has happened in the world. And by then, they'll be loads of new Facebook posts to look at. And then I haven't checked what people have put on Yammer or Instagram…

Procrastination is the thief of time. It's where we put off doing something that we should be doing in favour of doing something else that's more enjoyable or that you feel more comfortable doing. Sometimes, deadlines help focus the mind. But, as Douglas Adams, author of the *The Hitchhiker's Guide to the Galaxy* said: "I love deadlines. I love the whooshing noise they make as they go by".

Basically, your brain is divided up into two parts. There's the intellectual brain – cerebral cortex – which makes logical decisions and can be a little bit slower than the second part, the primitive brain – basically, the limbic system – which is more emotional and prone to act like a child (or even a monkey) and only wants to do, at any one time, what it wants to do. Your intellectual brain can overrule your primitive brain, but it takes effort. And, sometimes, only an impending deadline makes it worth making the effort.

So how do you know that you're procrastinating? The answer is that you probably do some or all of the following:

- Fill your day with low priority tasks.
- Read e-mails several times without starting work on them or deciding what you're going to do with them.
- Sit down to start a high-priority task, and almost immediately do something else (like check Facebook).
- Leave an item on your To-Do list for a long time, even though you know it's important.
- Regularly agree to do unimportant tasks and do those rather than getting on with the important tasks.
- Wait for the 'right mood' or the 'right time' to tackle important tasks.

So, what can you do to overcome your tendency to procrastinate? A recent Dutch study found that people were more likely to procrastinate when they were bored. And that fits nicely with the Lucille Ball quote: "If you want something done, ask a busy person to do it. The more things you do, the more you can do." Another study wondered whether people procrastinated because a task was unpleasant, or they were too disorganized to work out how much time the task required, or because they felt overwhelmed by the task and doubted their own skills to complete it.

So what techniques did the Dutch study come up with to help procrastinators? And these are techniques that you can use or you can pass on to your clients. Based on their PAWS (Procrastination At Work Scale) the study came up with 12 practical steps to implement change. They are:

1 If you delay making decisions, give yourself timescales. Ease yourself into these so that you don't set unrealistic targets but manage to set achievable ones that gradually help you perform more efficiently.

2 The delays you may make before starting tasks can have value if they ease you into the proper mindset. However, as with number 1 above, set limits on how long you give yourself to prepare before launching into the task.

3 Craving a diversion may reflect the fact that you're bored. Even with something mundane like raking leaves in the autumn, you can find ways to make it more of a challenge. How neatly can you pile those leaves, for example?

4 Daydreaming can be useful to an extent, but if it takes you out of the mindset you need to complete the task. Dig deeply into the task itself to find something of mental value, as in number 3 above. Alternatively, daydream while you complete the task, if it's one that's mindless enough.

5 Prioritizing is one of the best ways to avoid procrastination, but if you've got your priorities reversed, reward yourself for completing important (but difficult) tasks when you get them done on time.

6 Don't let the thought of a task that's too insurmountable keep you from starting it at all. Along with number 5 above, when you have a lot of work to do, break it down into manageable chunks.

7 Cutting down on breaks is easy if you have something to look forward to doing afterward. That break can be your reward instead of the distraction that prevents you from getting on with the task.

8 Mundane tasks can't be avoided entirely, but the more quickly you complete them, the more quickly you can get to do what you truly like.

9 Texting people when you've got something to do is an activity that you'll need to set limits on. Save your texting time for after you've finished what you need to do, or after you've finished a part of it.

10 If you've got the time, spend as much time as you want on Facebook, once the task is complete. Alternatively, use small periods of time on Facebook (or whatever) as a reward for completing part (a chunk) of the task.

11 It's easy to find yourself reading the news online, and losing track of how late it's getting. Reserve a certain part of your day for keeping up with the news, and stick to that schedule. You can also read the news online while you're having a coffee break.

12 Online shopping is potentially one of the most time-consuming ways of procrastinating. It generally takes longer than you expect. But if you're worried that the bargain you're after will disappear unless you buy it immediately, get it done and then go back to your important task. Online shopping can be used as reward for completing your task.

Other suggestions to prevent procrastination include:

- Do the worst task first thing, every day – the rest of the day can only get better!
- Then do some quick small tasks. You'll feel like you're achieving things.
- Keep a To-Do list so that you don't forget about big tasks.
- Identify the unpleasant consequences of *not* doing the task.
- Focus on one task at a time.

Knowing these techniques, it should be possible to overcome our own tendency to procrastinate. And we should be able to help our clients get on with their important tasks and not spend all day online, not actually making headway with those major tasks they complain that they never get done. The first task is to recognize that you (or they) are procrastinating.

But I mentioned in the introduction about being an above-average hypnotherapist. We can't all be above average can we? Suppose that the latest survey said that the majority of solution-focused hypnotherapist were able to perform above the average. Would you think that sounded right? Or would some small piece of a maths lesson from long ago keep nagging away at you that some people would be average and some people would be below average and only some would be left to be above average – not enough to be the majority. But is that right?

It's a bit like drivers really. If I were to ask a room full of people where they'd place themselves on a scale of driving ability, I'm pretty confident that the majority would place themselves well into the 'above average' section of the scale. And I would bet it's the same with hypnotherapists. All those years of experience, all those days spent successfully working with clients with quite difficult problems, and learning some arcane pieces of knowledge, I'm sure they would rate themselves well into the above average part of the scale. But does that make any sense? After all, the average has to be in the middle – everyone knows that!

Let's look at those car drivers for a moment. Suppose that we had a room with 100 car drivers in it. Let's suppose that we asked them to rate their driving ability on a scale of zero to 10. And let's suppose that 99 of them rated their driving ability as 10 out of 10. But, one person in the room kept having accidents in their car and was virtually uninsurable. They were feeling pretty glum today, and so they rated their driving skill as zero – and then went off to find someone to give them a lift home.

Now, let's do some maths. The mean average is calculated by adding up all the scores, taking the total, and dividing it by the number of scores used. That makes this calculation 99 times 10 divided by 100. There were 99 people who ranked themselves as having 10 points – giving a total 990 (plus one person with a zero score). And there were 100 people in the room taking the survey. So, the mean average works out at 990 divided by 100, or 9.9.

Therefore, as we've just seen, 99 of the 100 people in the room (the ones who rated themselves as 10 out of 10) are above-average drivers (because the average is 9.9).

And it's perfectly possible for the same result to be obtained with hypnotherapists. Many hypnotherapists have been working for a number of years and have seen a variety of clients – so they are definitely going to score themselves very highly (almost definitely 9s and 10s on the scale). There are also a few people who are just starting out as a hypnotherapist. For them, working with clients is still a bit scary and they often ask their supervisor for support. These people would realistically have to score themselves quite low (ones or twos on the scale).

For our drivers score, I took a fairly extreme example. However, the same result will occur wherever there is skewed data – by that I mean asymmetric data, ie more of the answers go with a particular value on the scale. So it is perfectly possible for, say 75% of hypnotherapists to be above average.

Those of you who know about maths, or who do your children's maths homework with them from time to time, will know that the mean isn't the only 'average'. And the reason that people don't think that more than half the people can be above average is because they mix it up with the median average. The median average is the number in the middle of a list of numbers after they have been sorted into numerical order. If there is no middle digit, then it is calculated as the mean average of the middle two digits. 50 percent of the results are higher than the median average and 50 percent lower. But we have used the mean average here.

Although no-one has surveyed hypnotherapists to see how good they think they are, I'm sure you'll agree with me in being confident that the majority of them are well above the average!

References:

https://www.psychologytoday.com/blog/fulfillment-any-age/201611/12-ways-get-over-procrastination-now

https://www.mindtools.com/pages/article/newHTE_96.htm

https://en.wikipedia.org/wiki/Procrastination

http://waitbutwhy.com/2013/10/why-procrastinators-procrastinate.html

Metin, U. B., Taris, T. W., & Peeters, M. W. (2016). Measuring procrastination at work and its associated workplace aspects. Personality and Individual Differences, 101254-263. doi:10.1016/j.paid.2016.06.006 – http://www.sciencedirect.com/science/article/pii/S0191886916307474

Clarry H Lay (1968). At Last, My Research Article on Procrastination. *Journal of Research on Personality* – http://www.sciencedirect.com/science/article/pii/0092656686901273

Using Appreciative Inquiry in a therapy situation

Here's how to use Appreciative Inquiry with clients.

Appreciative Inquiry (AI) starts by looking at the best of what currently exists, in order to imagine what could be, followed by designing a desired future state that is so compelling that people want to move towards it. Rather than asking what's wrong (problem focused), it starts by asking what's right (which is more solution-focused).

An example question to ask your client would be: "Tell me about the best part of your day yesterday?"

Not only does it focus their attention on positives rather than the more usual negatives, clients may even notice a difference in how they felt when considering the answer. Some people experience a greater sense of calm or joy, than if they were only recalling what happened during the day. Savouring good experiences is, apparently, one key to happiness.

The 4D model of AI was developed by Cooperrider and Whitney in 2005. Its four steps are Discovery, Dream, Design, and Destiny.

During the Discovery stage, the client should be asked questions identifying and detailing their 'positive core', including information about their family and work. This is meant to generate a dialogue about the "root causes of [their] success".

Some questions to ask include:

- What do you think is your core factor or value, which you wouldn't be the same without?
- If you had three wishes for yourself, what would they be?
- What achievements are you proud of?
- Who are the important people in your life?
- What have been your best experiences with…?
- What do you value about… yourself, your relationship with…
- In what situations is the problem most/least likely to occur?
- How have you attempted to solve the problem?
- What other strategies might you try?
- How would you know that you had been successful?

In the Dream stage, clients build on what's good already, the positive resources they have, and any new ideas that come out of the Discovery stage to create a vision of how they will be in the future.

The Design stage takes the vision from the Dream stage and works out the steps needed to get from here to there. This may involve challenging the *status quo*. It helps the client to have a clear idea of their goal.

The Destiny stage is all about implementing the steps identified in the design stage. The client becomes responsible for taking the steps to achieve the Dream (goal). Their decisions and actions must move them towards their Dream.

Our job is to help them to take the necessary steps.

References:

https://en.wikipedia.org/wiki/Appreciative_inquiry

https://www.psychologytoday.com/blog/life-changes/201001/how-coaching-works-appreciative-inquiry

http://familytherapybasics.com/blog/2017/3/13/appreciative-inquiry-how-to-facilitate-powerfully-positive-change-in-therapy

http://coachingleaders.co.uk/what-is-appreciative-inquiry/

Vegan therapists

I always picture hypnotherapists drinking high-caffeine fizzy drinks and munching their way through extra-large burgers. But it seems that many hypnotherapists are quietly becoming vegans for their health. What reasons are some people giving for cutting out meat from their diet?

There seem to be basically three reasons people give for cutting out meat from their diet – the conditions that animals are kept in, the chemicals that are given to the animals that are then eaten by people, and not wanting to eat other animals.

Let's look at what people say happens to chickens. Chickens raised for their flesh are called 'broilers', and spend their entire lives in sheds with thousands of other birds. This intense crowding and confinement leads to outbreaks of disease. They're drugged to grow larger than normal, which leads to heart attacks, organ failure, and leg deformities. At 6 or 7 weeks old, they're sent to slaughter. Chickens raised for their eggs are called 'laying hens', and are crammed together inside wire cages with a floor space of 25cm by 22cm. Hens often can't flap their wings or stand upright. Careful manipulation of light and the amount of food the birds receive increases egg size and production. The chickens urinate and defecate on one another, and have part of their beak cut off to stop them pecking each other. When they stop laying, they are slaughtered, usually for chicken soup or cat or dog food. Male chicks are killed shortly after hatching. At the slaughterhouse, chickens are forced into shackles, have their throats cut, and are immersed in hot water to remove their feathers.

> Hens naturally scout their environment, forage and peck around, determine social hierarchies, build nests, and groom themselves.

There's nothing wrong with beef, is there? Apparently, cows are often stuffed into Confined Animal Feed Operations (CAFOs) and given a diet primarily of genetically-modified maize and grains. The cows are forced to stand in manure and urine while feeding or being milked. They're given antibiotics to prevent disease and also to stimulate growth. Steroid hormones, including natural oestrogen, progesterone, and testosterone are given to cattle. These drugs increase the animals' growth rate and the efficiency with which they can convert the food they eat into meat. Synthetic hormones, like trenbolone acetate, progestin, melengestrol acetate, and zeranol, are also used to make animals grow faster and/or produce leaner meat for food. The drug ractopamine is fed to cattle, pigs, and turkeys to make them produce larger quantities of leaner meat with less feed. Recombinant Bovine Growth Hormone (rBGH) is injected into cows to increase milk production.

What about what happens to pigs? Pigs, it seems, are kept in small cages so they are unable to turn round. Breeding sows are kept in gestation crates, which are only slightly larger than their body. The crate floors are usually made of slats, which allow manure to fall through, meaning that sows live directly above their own waste. The flooring causes excessive foot injuries, damage to joints, and even lameness in the pigs. Just before any piglets are born, the sows are moved to 'farrowing crates'. At 17–20 days old, the piglets are taken away from their mothers and are castrated and have their tails docked. The piglets are confined for the next 6 months until they reach 'market weight' and are sent to slaughter. Pig farmers use nearly four times as much antibiotic as cattle farmers do, per pound of meat produced.

Are sheep OK – don't they just graze on the hillside? Lambs are usually born in the spring, but many farmers choose to lamb during the winter, so the lambs are big enough to slaughter for the 'spring lamb' market around Easter time. They achieve this by using drugs and hormone implants to manipulate the sheep breeding cycle. After birth, lambs are castrated and have their tails docked. Drugs for a wide range of external and internal parasites are administered by injection, poured down the animal's throat, or applied through whole-body immersion. Sheep dipping prevents scab and blowflies.

The argument being made here is that, in many ways, there is nothing wrong with meat, milk, and eggs, it's just the way that they are produced these days. Many hypnotherapists care about their health and are less keen to fill their bodies with hormones, antibiotics, and other chemicals. And, also, they care about the welfare of animals and don't want to encourage this kind of treatment. And, thirdly, they don't want to eat other mammals (or birds).

There are health reasons to cut out meat from your diet. It's suggested that you have a lower risk of cancer (the Physicians Committee for Responsible Medicine reported that vegetarians are less likely to get cancer by 25 to 50 percent) and heart disease (Drs Dean Ornish and Caldwell Esselstyn have a program that includes a vegetarian diet and is currently one of the few programs that has been proven to reverse heart disease and reduce cholesterol levels). And there's a lower risk of osteoporosis (too much protein in our diet causes loss of bone calcium) and kidney and gall stones.

Anecdotally, I know quite a few people who are moving towards or have already become vegetarians or vegans. Some are pescatarians (they only eat wild fish and seafood with their fruit and vegetables). Or they only eat meat that comes from old-fashioned farms where the animals are allowed to wander round and forage for their own food.

What's your experience?

References:

http://www.animalliberationfront.com/Practical/Health/15Reasons2NotEatMeat.htm

http://www.aspca.org/animal-cruelty/farm-animal-welfare

Yuval Noah Harari. Sapiens: A Brief History of Humankind. Vintage; 978-0099590088

Staying legal

A brief look at the impact of new regulations on small hypnotherapy businesses.

The European General Data Protection Regulation (GDPR) comes into force on 25 May 2018, but what does it mean for your business? Every company will have to comply with new regulations affecting the secure collection, storage, and usage of personal information – with fines for violations.

The whole point behind GDPR is to:

- give back control of their personal data to citizens and residents; and
- simplify the regulatory environment for international business by unifying the regulation within the EU.

On the plus side, the GDPR recognizes that smaller businesses require different treatment to large ones. Article 30 states that organizations with fewer than 250 employees will not be bound by GDPR!

But, the GDPR applies to small businesses with under 250 employees if the processing carried out is likely to result in a risk to the rights and freedoms of data subjects, the processing is not occasional, or the processing includes special categories of data as defined in GDPR Article 9.

Be aware, that the GDPR says:

- Breaches in data security must be reported immediately (in 24-72 hours) to data protection authorities such as the Information Commissioner's Office (ICO) in the UK.
- Individuals have more rights dictating how businesses use their personal data. In particular, they have the 'right to be forgotten' if they either withdraw their consent to the use of their personal data or if keeping that data is no longer required.

The Information Commissioner's Office (ICO) is an independent regulatory office dealing with the Data Protection Act 1998 and the Privacy and Electronic Communications (EC Directive) Regulations Its mission is to "uphold information rights in the public interest, promoting openness by public bodies and data privacy for individuals". It can issue fines, known as monetary penalties, up to £500,000 for breaches of the Privacy and Electronic Communications Regulations (PECR). PECR applies to organizations that wish to send marketing messages through electronic means, ie phone, fax, e-mail, text; use cookies; or provide electronic communication services to the general public. And that's where it can apply to hypnotherapists if you do any of those things. You can check whether you need to register by taking their assessment at https://ico.org.uk/for-organisations/register/self-assessment/. It costs £35 to join unless you're a big organization, when it costs more.

- Failure to comply with the GDPR means the Information Commissioner's Office (ICO) can fine up to £500,000 for malpractice. However, the GDPR will be able to fine up to €20 million or 4 per cent of annual turnover (whichever is higher).

Companies will not only have to get the clear and unambiguous consent of their customers to store and use their personal data, they will also have to keep a secure record of how and when that consent was granted, what it was granted for, and for how long. You must be able to produce a clear audit trail of consent.

GDPR applies to you if you regularly deal with personal data, including present and past employees, suppliers, and customers. If you do, you should comply with the GDPR. The ICO says that any businesses affected by the DPA (Data Protection Act) will also fall under the GDPR.

All data should be stored securely and processes should be in place to ensure personal data is kept separately under a security framework.

You may also get Subject Access Requests (SARs). Under the Data Protection Act, these are requests by individuals to see a copy of the information an organization holds about them. And you may get the 'right to be forgotten', which requires you to identify and erase all of an individual's data.

Step 1 for implementing GDPR may require an information audit, so you understand exactly what information is stored, where, and for what purpose. Step 2 is to organize that data and ensure it is protected and can't be accessed by unauthorised people or applications. You should be thinking about password protecting it and perhaps encrypting it.

It's worth getting organized now before the new regulations come into effect.

About the author

Trevor Eddolls BA, Cert Ed, MOS MI, DHP, HPD, SFBT Sup (Hyp), CBT (Hyp), Dip NLP, Dip Mindfulness, AfSFH (Exec), CNHC Registered, UKCHO Registered is a clinical hypnotherapist and psychotherapist. He's clinical director at iTech-Ed Hypnotherapy and Head of IT on the AfSFH (Association for Solution-Focused Hypnotherapy) Executive. Trevor is a Hypnotherapy Master Practitioner, a Solution-Focused Hypnotherapy Supervisor, and an NLP Master Practitioner. He has a diploma in CBT and a diploma in Mindfulness. He is a qualified Life Coach, and has a diploma in nutrition and a diploma in paediatric hypnotherapy (Dip Hyp (paediatrics)).

Solution-focused hypnotherapy, as its name suggests, focuses a client's attention on the solution to their problems rather than the causes. Evidence suggests that dwelling on what led to a problem can increase the client's issues, whereas focusing on solutions can dramatically reduce those issues.

Trevor is a popular blogger and presenter. He has been seeing clients and writing about hypnotherapy, CBT (Cognitive Behavioural Therapy), NLP (Neuro-Linguistic Programming), and Mindfulness techniques for around 10 years.

Before training as a hypnotherapist, Trevor worked with mainframes. He also spent many years writing books and articles, and editing well-respected technical journals about mainframe technology.

You can contact Trevor at iTech-Ed Hypnotherapy in the Wiltshire town of Chippenham.

His Web site is at www.ihypno.biz.

Facebook: facebook.com/iHypno2004

Twitter: twitter.com/iHypno2004

Instagram: instagram.com/ihypno2004